Applications of Plasma Source Mass Spectrometry II

Applications of Plasma Source Mass Spectrometry II

Edited by

Grenville Holland
Department of Geological Sciences, University of Durham

Andrew N. Eaton
VG Masslab, Wythenshawe, Manchester

ROYAL
SOCIETY OF
CHEMISTRY

The Proceedings of the 3rd International Conference on Plasma Source Mass Spectrometry held at Durham, UK on 13–18 September, 1992.

Special Publication No. 124

ISBN 0-85186-465-1

A catalogue record for this book is available from the British Library

Published by The Royal Society of Chemistry,
Thomas Graham House, Science Park, Cambridge
CB4 4WF

Printed in Great Britain by Bookcraft (Bath) Ltd.

Preface

The rapid development of inductively coupled plasma source mass spectrometry and its associated techniques has provided the most important analytical system in the last decade of this century. Every year this progress is charted in analytical conferences all over the world, non more so than at the Durham International Conference on Plasma Source Mass Spectrometry. The proceedings of the first of these conferences went unrecorded; but the second conference provided the book entitled "Applications of Plasma Source Mass Spectrometry". The third conference, held in September 1992, was not only immensely enjoyable but has also generated a second volume to add to the first.

This second volume, slightly larger than the first, provides a faithful record of the papers and posters presented in Durham last September. An examination of the contents page demonstrates the breadth of topics discussed at the conference and measures the level of development of plasma source mass spectrometry at that moment in time. Although all the papers contained in this book have been refereed and checked, as in the first edition we have allowed the authors' ideas and perception of the science to have free rein. These papers therefore represent, with as much fidelity as possible, the views of the authors and their associates.

A conference as enjoyable socially and scientifically as the Durham meeting requires the help and support of many people, not least the delegates. However, we owe a particular debt to the many staff at VG Elemental, most especially Dr. Robert Hutton and Miss Karen Morton, whose unfailing support and encouragement ensures that this particular conference is such a success. In Durham, Mrs. Beatrice Smith ran the administration with her usual friendly courtesy and efficiency and dealt calmly with the idiosyncracies of our scientific community. Finally our special thanks to the staff of the Royal Society of Chemistry for their patient good nature.

Grenville Holland
University of Durham

Andrew Eaton
VG MassLab

Contents

Determination of the First Transition Elements in Terrestrial Water by High Resolution ICP-MS with an Ultrasonic Nebulizer

S. Yamasaki and A. Tsumura
NATIONAL INSTITUTE OF AGRO-ENVIRONMENTAL SCIENCES, 3-1-1
KANNONDAI, TSUKUBA, IBARAKI, JAPAN 305

T. Kobayashi
KITAZATO UNIVERSITY, SAGAMIHARA, JAPAN 228

1 INTRODUCTION

In recent years, there has been an ever increasing concern about the concentrations of the first transition elements ([21]Sc to [30]Zn in atomic number) in environmental samples. In spite of this growing interest, however, there appears to be insufficient data, at present, about the contents of these elements in terrestrial water because the concentration levels of the first transition elements, especially those of Sc, Ti, and Co, in water samples are very low in most cases.

Although inductively coupled plasma mass spectrometry (ICP-MS) has been recognized as the most sensitive technique for elemental analysis, attempts to determine trace levels of the first transition elements in water samples by the commonly used ICP-MS often encounter difficulties. This is mainly because quadrupole-type mass spectrometers are incapable of separating various overlapping polyatomic peaks due to gaseous components (H, C, N, O, and Ar) and/or matrix elements in water samples (Na, Ca, S, and Cl). In addition, the concentration levels of most of the first transition elements in water samples are usually too low to allow direct determination even with this extremely sensitive method.

A high resolution ICP-MS with a double-focusing type of mass spectrometer has been developed to overcome the problems of spectral overlaps [1-2]. Meanwhile, it has been pointed out that the use of an ultrasonic nebulizer (USN) significantly improves the sample introduction efficiency, and has resulted in greatly increased sensitivity in inductively coupled plasma atomic emission spectrometry (ICP-AES) [3-5]. In order to obtain an analyzing system with greater detection power, we have attempted to combine the high resolution ICP-MS with a USN, and have examined its capabilities by analyzing trace and ultra-trace levels of the first transition elements in water samples.

2 EXPERIMENTAL

Instrumentation

High Resolution ICP-MS. The measurements were carried out using a double-focusing type of ICP-MS (PlasmaTrace) with much higher resolution supplied by VG Elemental, Winsford, Cheshire, England. This instrument is capable of operating at resolutions (M/ΔM) of up to 10,000. The details of the system have been described previously [6-7]. The typical operating conditions are given in Table 1. Selected isotopes were basically for those of highest abundances. But a less abundant isotope was chosen for Ti and Ni to avoid possible isobaric interference respectively due to Ca and Fe.

Table 1 Operating Conditions for High Resolution ICP-MS

Inductively Coupled Plasma

Forward Power	1.2	kW
Reflected Power	< 2	W
Coolant Flow Rate	14	L/min
Auxiliary Flow Rate	1.2	L/min
Nebulizer Flow Rate	1.1	L/min

Mass Spectrometer

Accelerating Voltage	4	kV
Sample Orifice (Nickel)	1	mm
Skimmer Orifice (Nickel)	0.5	mm
Resolution	3000 - 3500	

Data Acquisition

Dwell Time	160 ms
No. of Points	70
DAC Step	10
No. of Scans	1

Isotopes Selected

^{45}Sc ^{47}Ti ^{51}V ^{52}Cr ^{55}Mn
^{56}Fe ^{59}Co ^{60}Ni ^{63}Cu ^{64}Zn

Ultrasonic Nebulizer. The ultrasonic nebulizer used in this study was provided by Applied Research Laboratories, En Vallaire, CH-1024, Ecublens, Switzerland. A schematic diagram of the device is shown in Figure 1. A peristaltic pump is used to deliver sample solution to the oscillating surface of the quartz plate. Aerosols having smaller and more uniform particle size are produced, and transported first to the heated tube, and then to the condenser by the Ar gas for the nebulizer. Most of the water in the aerosol is removed by this process, and only the dry aerosol is introduced into the plasma. The temperature of the heated tube and the condenser was kept at 120°C and 1°C respectively. The sample introduction rate was adjusted to 3 ml/min.

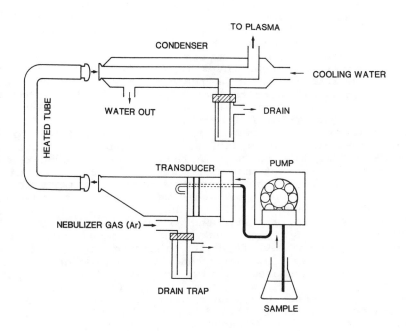

Figure 1   Schematic Diagram of the USN

Reagents

"Ultra-pure water," prepared for the semiconductor industry, was used in the experiments. Samples and standard solutions were acidified by the ultra-high purity nitric acid (Tamapure-100) provided by Tama Chemical Industry Co., Ltd., Tokyo, Japan. The contents of various metals in the acid were guaranteed to be less than 100 ppt. The working standards (0, 25, 50 and 100 ppt) were prepared from a series of SPEX Multi-Element Plasma Standards supplied by SPEX Industries, Inc., 3880 Park Ave., Edison, New Jersey, U. S. A.

Sample Handling

Water samples were collected in specially prepared polyethylene bottles. The bottles were made from metal-free raw materials in a class 100 clean room without using plasticizer. The bottles thus made were thoroughly washed, also in a clean room, with ultra-pure water. The collected samples were kept in a refrigerator at about 5°C until needed for determination by ICP-MS. The

samples thus stored were then filtered through a 0.45 μm
cellulose acetate filter just before the measurements.

3   RESULTS AND DISCUSSION

Typical Mass Spectra

Figure 2 summarizes the mass spectrum obtained by
the standard solution containing 1 - 100 ppt of the
first transition elements.   It can be seen that there
were molecular peaks at a nominal mass of 45, 51, 52,
55, and 56, which were respectively identified as $^{28}Si^{16}OH$
+ $^{29}Si^{16}O$ and $^{12}C^{16}O_2H$ + $^{13}C^{16}O_2$, $^{35}Cl^{16}O$, $^{40}Ar^{12}C$ +
$^{36}Ar^{16}O$, $^{40}Ar^{14}NH$ and $^{40}Ar^{16}O$.   The origin of Si for
molecular peaks at nominal mass 45 is considered to be
from the Si in the glassware used for sample introduction
system and/or the torch.   Although ion chromatographic
analysis proved that the content of Cl in the standard
solution was less than 100 ppt, the peak at mass 51 was
identified as ClO because other peaks were also observed
at a nominal mass of 35, 37, and 53 ($^{37}Cl^{16}O$).   It was
not possible to specify the origin of this Cl.   The
signal intensity of the peak at mass 52 was much higher
than that calculated from the abundance ratio of ^{36}Ar
to ^{40}Ar.   It was also observed that the height of the
peak sharply increased when solutions containing organic
matter were introduced.   From the foregoing results, it
can be concluded that this peak is ascribed to the
overlapped peaks of $^{40}Ar^{12}C$ and $^{36}Ar^{16}O$.   The needed
resolution for the separation of these two molecular
species is calculated to be more than 680,000 and,
hence completely beyond the ability of the instrument.

Figure 3 illustrates a series of mass spectra
respectively obtained by 10 ppt of Co and Cu with 10 ppm
of Na, and those by Co and Ni with 10 ppm of Ca.   As
Figure 2 shows, there appears to be no interfering peak
for those elements when the standard solutions containing
only the elements of interest   were introduced.   The
occurrence of several new molecular species was clearly
observed, however, at nominal mass 59, 63, 59, and 60
if Na or Ca was added as a matrix element.   These peaks
were respectively ascribed to $^{36}Ar^{23}Na$, $^{40}Ar^{23}Na$,
$^{43}Ca^{16}O$, and $^{44}Ca^{16}O$.   As the concentration levels of
the added Na and Ca are typical or somewhat lower than
those of real river water samples, it is obvious that
the determination of trace amounts of these elements is
deemed to be inaccurate if the separation of overlapping
peaks is not complete.

In addition to those molecular species mentioned
above, there were several other molecular species due
to Li and S at nominal mass 47 and 64 in real water
samples.   Furthermore, the peak heights of $^{28}Si^{16}OH$ +
$^{29}Si^{16}O$ and $^{35}Cl^{16}O$ were much higher in freshwater samples

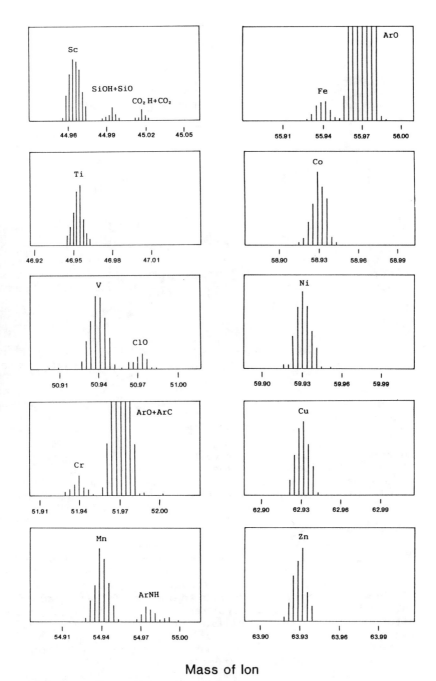

Mass of Ion

<u>Figure 2</u> Mass Spectra of the First Transition Elements

See Table 2 for the concentration of each element.

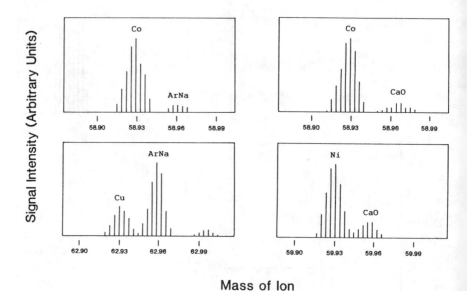

Figure 3 Mass Spectra of Co and Cu with Na (left)
and Co and Ni with Ca (right)

The concentation is 10 ppt for Co, Cu, and Ni, and 10 ppm
for Na and Ca.

because the concentration levels of Si and Cl were usually
10 - 100 ppm in most cases. It is reasonable, therefore,
to assume that accurate and precise determination of ppt
levels of the first transition elements in water samples
by quadrupole-type machines are most unlikely.

Resolution versus Sensitivity.

Figure 4 depicts how the signal intensity changes
as the resolution increases. Data were obtained using
1 ppb of indium (In) solution. Signal intensity was
expressed in terms of counts per second (Hz) per ppm of
In. The analyzing system used in this experiment
attained such an excellent detection power that the
sensitivity reached as high as 2 GHz/ppm when the
instrument was operated in the low resolution mode
(around 400). When the resolution was increased to 1500
to separate, for example, Si from N_2, the sensitivity
was decreased to 60% of that obtained in the low
resolution mode. If the resolution was further increased
to 3500 for the first transition elements, the signal
intensity was around 400 MHz/ppm. At a resolution of
8500, the sensitivity was only 2% of that attained in
the low resolution mode.

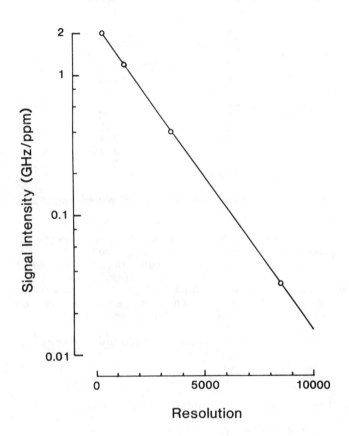

Figure 4   Sensitivity versus Resolution Curve

Typical Detection Limit

    The detection limits of the first transition elements
obtained by the PlasmaTrace/USN combination are listed
in Table 2, together with the integrated counts of the
standard (1-100 ppt) and the blank solution. Detection
limits are defined here as the equivalent concentration
of three times the standard deviation of the blank
response.   The standard deviation of the blank solution
(also shown in Table 2) was calculated from 10 consecutive
measurements carried out within a relatively short
period (usually less than 3 min).   The total peak dwell
times were about 5 sec.

Table 2   Typical Detection Limit

Element	Standard Conc. (ppt)	Standard Total Counts*	Blank Total Counts*	Detection Limits (ppt)
Sc	1	393	8.20 ± 2.50	0.019
Ti	100	98	0.14 ± 0.31	0.95
V	10	164	0.14 ± 0.14	0.026
Cr	10	397	0.57 ± 1.16	0.088
Mn	10	414	4.11 ± 1.36	0.098
Fe	10	4099	370 ± 285	2.09
Co	10	183	0.86 ± 0.46	0.075
Ni	10	318	5.00 ± 2.16	0.20
Cu	10	662	4.11 ± 1.83	0.083
Zn	10	920	5.25 ± 1.71	0.056

* Average and standard deviation of 10 measurements

It can be seen that the detection limits of the first transition elements are all well below 1 ppt except for Fe. Whereas the high value for Fe can be attributed to the higher total counts of the blank presumably due to the effect of ArO, somewhat higher detection limit of Ti is the consequence of the low abundance of [47]Ti used in this experiment (7.32%).

Analytical Results for a Standard Reference Material

The usefulness and reliability of the proposed methods were tested using the Standard Reference Material SRM 1643c (Trace Elements in Water) provided by the National Institute of Standards and Technology [8]. This reference material was analyzed after a 100-fold dilution because the concentration levels of elements in the sample were much higher than the optimum level of the system used here. Sodium and Ca were also added to bring the final concentration of these elements to 10 ppm as matrix components.

Table 3   Analytical Results of SRM 1643c

Element	Current Study (ppb)*	Certified Value (ppb)
V	30.5 ± 2.2	31.4 ± 2.8
Cr	20.6 ± 1.1	19.0 ± 0.6
Mn	35.5 ± 1.4	35.1 ± 2.2
Fe	112.1 ± 5.3	106.9 ± 3.0
Co	23.1 ± 1.7	23.5 ± 0.8
Ni	53.5 ± 4.9	60.6 ± 7.3
Cu	23.1 ± 2.1	22.3 ± 2.8
Zn	72.9 ± 6.0	73.9 ± 0.9

* Average and standard deviation of 5 measurements

The results summarized in Table 3 demonstrate that values obtained in this work were generally in good agreement with the certified values even though the concentration levels in the final solution were less than 1 ppb, with the exception of Fe.

Mean and Range of the First Transition Elements

The arithmetic mean and range of the concentration of the first transition elements in lake and river water, randomly sampled at various places in Japan, as well as several samples collected in Argentina, England, Indonesia, Thailand, and Russian Federation are shown in Figure 5.

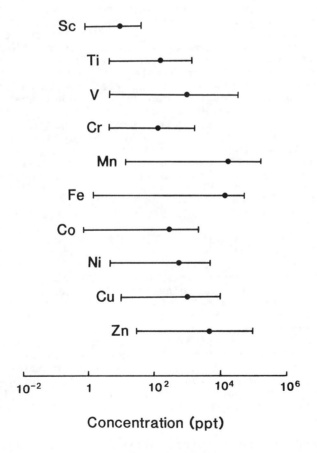

Concentration (ppt)

Figure 5   First Transition Elements in Water

Bars and dots show the concentration ranges and the arithmetic means respectively.

The concentration ranges were so wide that the ratios of the highest values to the lowest values are more than 1000 for all the elements shown in Figure 5, with the exception of Sc. Whereas the concentration levels of Mn, Fe, and Zn tended to be higher than the other elements, the reverse was the case for Sc, Ti, and Co. Although the maximum concentration of some elements was higher than 100 ppb, the percentage of samples that contained more than 1 ppb of the first transition elements was less than 50%, except for Fe (54%) and Zn (78%).

4 SUMMARY

A specially designed, high resolution ICP-MS was combined with an ultrasonic nebulizer to analyze trace and ultra-trace amounts of the first transition elements (^{21}Sc through ^{30}Zn in atomic number) in freshwater samples. The occurrence of a large number of polyatomic molecular species was confirmed in the mass range of the first transition elements (45 - 64). From the comparison of the signal intensity of the interferents and that of the analytes, it can be concluded that accurate and precise determination of ppt levels of the first transition elements in water samples by quadrupole-type ICP-MS are most unlikely. The resolution needed to separate those interferents from the analyzing elements is around 3000. Although there is a considerable decrease in the sensitivity when the resolution of the instrument was increased to 3000, the proposed method was so sensitive that the detection limits of all the above elements were still well under 1 ppt, except for Fe.

Analytical results for the Standard Reference Material 1643c (trace elements in water) obtained in this study were in good agreement with certified values for all the elements examined. The concentrations of the first transition elements in diverse freshwater samples ranged so widely that the ratios of the highest values to the lowest values were more than 1000 for all the elements, with the exception of Sc. Whereas the concentration levels of Mn, Fe, and Zn tended to be higher than the other elements, the reverse was the case for Sc, Ti, and Co. Although the maximum concentration of some elements was higher than 100 ppb, the percentage of samples that contained more than 1 ppb of the first transition elements was less than 50% except for Fe and Zn.

Based on the results obtained, it can be concluded that the PlasmaTrace/USN combination provides such an extraordinary sensitive method that accurate and precise determination of ppt levels of the first transition elements in water samples are easily achievable.

ACKNOWLEDGEMENT

The authors would like to thank Hashimoto Co., Ltd., Osaka, Japan for supplying ultra-pure water. Thanks are also due to F. Ochida for assistance in preparing the figures.

REFERENCE

1. N. Bradshaw, E. F. H. Hall and N. E. Sanderson, J. Anal. At. Spectrom., 1989, 4, 801.
2. M. Morita, H. Ito, T. Uehiro and K. Otsuka, Anal. Sci., 1989, 5, 609.
3. K. W. Olsen, W. J. Haas, Jr., and V. A. Fassel, Anal. Chem., 1977, 49, 832.
4. V. A. Fassel and B. R. Bear, Spectrochim. Acta, 1986, 41B, 1089
5. R. J. Thomas and C. Anderau, Atom. Spectrosc, 1989, 10, 71.
6. A. Tsumura and S. Yamasaki, 'Applications of Plasma Source Mass Spectrometry', The Royal Society of Chemistry, Cambridge, 1991, p. 120
7. S. Yamasaki and A. Tsumura, Anal. Sci., 1991, 7 (Sup.), 1135.
8. National Institute of Standards and Technology, 'Certificate, Standard Reference Material 1643c.' 1991

Determination of Trace Metal Impurities in Electronic Chemicals by ICP-MS and Other Methods

B. Gercken, J. Pavel, and O. Suter

CENTRAL ANALYTICAL DEPARTMENT, CIBA-GEIGY LTD., CH-4002
BASLE, SWITZERLAND

1 INTRODUCTION

The requirements on lowering metal impurities in electronic chemicals are becoming more and more stringent due to
the increasing memory capacity of the integrated circuits.
Critical metal impurities are: Alkali- and alkaline earth-
metals which may cause corrosion and degradation of the
microelectronic devices, heavy metals may increase the
resistance and may build up potentials on the surface and
alpha-emitters U and Th may cause so-called *soft errors* in
the storage devices[1]. Alpha-particles ionize the atoms they
pass during their flight. If an alpha-particle hits and
penetrates an uncharged memory cell, this cell will get
filled up with electrons. In the binary system, this means
that a ONE will get transformed to a ZERO. This is the way
soft errors cause inversion of the stored information.

The manufacturer of materials for microelectronic industry specify their products according to the requirements[2]. Suggested guidelines and standards exist both for
the manufacture and analytical methods[3]. Besides the production, which has to be done exclusively in clean rooms to
produce as contamination-free as possible, the analytical
chemistry is becoming an essential part of the production.
This means the analytical techniques have to cope with the
demands on the purity in the range of ultra trace elements.

Our experience of analyses of electronic chemicals includes determination of trace metal impurities in polyimide
and Novolak photoresist products as well as raw materials,
raw products and organic solvents. This presentation will
focus on analyses of polyimides as a comprehensive example
of our efforts on this topic. Polyimides are produced by
the Polymer Division of Ciba-Geigy Ltd. Formulated polyimide products are applied as photosensitive lacquers to
silicon wafers. Besides masking applications of photosensitive versions of some products, polyimides offer some
other useful properties in terms of protection of the wafer
such as radiation resistance and thermal, mechanical and
chemical stability[4,5].

Analytical Methods and Sample Preparation Techniques

ICP-MS is a very capable analytical method in terms of sensitivity and detection limits for most metals. This is also valid for Zeeman Graphite Furnace Atomic Absorption Spectrometry (ZGFAAS) though there is the disadvantage of not being capable of simultaneous multielement analysis. Limitations for the determination of trace metal impurities in electronic chemicals do occur due to sample preparation. Blank values of the entire procedure have to be kept as low as possible because they often represent the limiting factor. For this reason sample preparation should be as simple as possible to avoid losses as well as contaminations.

There are several options for sample preparation of polyimides that can be considered:

- Digestion of the solid polyimide sample in a PTFE-bomb with acid. This method is well established but means strong dilution of the original sample. Loss of sensitivity as well as contamination may result.

- Dry ashing of the sample involving cool plasma ashing prior to measurement. The procedure is rather time-consuming but the advantage is that due to high sample weight of 2 - 5 g the dilution factor is small (<5). This helps to avoid contaminations and to achieve low detection limits.

- Direct analysis of solid polyimide product after dissolving in an organic solvent. This seems to be a rather simple method but the requirements concerning the purity of the organic solvent are very high. Another critical point is that the trace metal impurities might not be dissolved completely. And last but not least: direct usage of organic solvents in plasma spectrometry does need special precautions, dedicated calibration and is difficult from the practical point of view as changing between water and organic solvent has to be done regularly.

- Direct analysis of the polymer product involving laser ablation technique or electrothermal vapourization (ETV). This option seems easily achievable and convenient. Danger of contaminating the sample material disappears but the quantitation seems rather difficult. The practical experience we have on this topic is very small .

2 EXPERIMENTAL

ICP-MS

The method of our choice involves cool plasma ashing (CPA) of the sample material prior to ICP-MS measurement. The method was originally developed for the determination of U and Th[6,7]. Figure 1 shows sample preparation steps. 2 - 3 g of solid polyimide or 4 - 5 g of formulated polyimide (dissolved in N-Methylpyrrolidone) were dry ashed by a CPA-4 system (operating conditions are listed in Table 1). Remaining residue was dissolved by refluxing with 0.4 ml H_2O, 0.4 ml HNO_3 and 0.2 ml HCl at 130°C for 1.5 h. Solutions were diluted to 10 ml and 100 ppt [233]U were added as internal standard for U- and Th-determination in *peak jump* mode. All other metals were measured involving scanning mode after addition of 100 ppb In as internal standard.

In order to verify this procedure, recovery experiments were performed involving NIST Standard Reference Material (SRM) 1634b *Trace Elements in Fuel Oil* (100 mg) as well as elemental standards of U and Th and multielement standards (ICP-mulitelement standard solution IV, Merck and N930-0211 and N930-0213 PE Pure Multi-Element Standards, Perkin Elmer) (50 mg) containing other metals of interest (Al, Ba, Ca, Cu, Cr, Fe, K, Mg, Mn, Na, Ni, Pb, Zn).

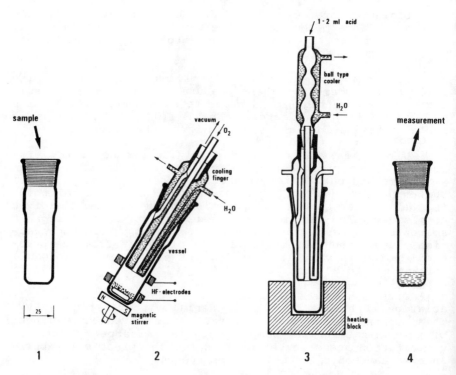

Figure 1: Cool plasma ashing operation scheme[8]; sample preparation steps: sampling (1), ashing (2), refluxing (3) and measurement (4)

ZGFAAS (Zeeman Graphite Furnace Atomic Absorption Spectrometry)

Solid samples were mineralized by plasma-ashing as described under ICP-MS or by heating to 160°C for 5 - 6 h in a

PTFE-lined pressure bomb (50 mg, 0.4 ml HNO_3 + 0.1 ml HCl).
Cu, Fe and Na were analyzed from obtained solutions with a
Perkin-Elmer Z-5100 Zeeman graphite furnace atomic absorption spectrometer.   External standard method was used for
calibration.

Table 1: Operating conditions for analysis of U und Th
         (*Peak Jump* Mode) and other metals (*Scanning* Mode)
         in Polyimides via Cool Plasma Ashing and ICP-MS

<u>Cool Plasma Ashing</u>   CPA-4 (Anton Paar KG., Graz, Austria)

RF-Power / W	40
O_2-Flow / l min^{-1}	2
Ashing time / h	14 - 18

<u>ICP-MS</u>   VG PlasmaQuad 1 (Fisons Instruments, Winsford, UK)

RF-Power / W	1300
Reflected Power / W	<20
Nebulizer type	V-Grove
Spray chamber	Scott Type
Torch	Fassel
Plasma gas flow rate / l min^{-1}	12.5
Auxiliary gas flow rate / l min^{-1}	1.0
Nebulizer gas flow rate / l min^{-1}	0.7
Sample uptake rate / ml min^{-1}	1.0

Sanning Parameters (analysis of metals)

Range / amu	5 - 242
Channels	2048
Dwell Time/channel / μs	250
Sweeps	100

Peak Jump Parameters (analysis of U and Th)

masses selected	232, 233, 238
Points/Peak	5
Dac Steps	5
Dwell Time / μs	10240
Sweeps	100

<u>ICP-OES (Inductively Coupled Plasma-Optical Emission Spec-
trometry</u>

Ca was measured using a *Philips PV 8065/90* Plasma
Optical Emission Spectrometer providing simultaneous
measurements with an automatic background accessory (at
396.85 nm).

XRFS (X-Ray Fluorescence Spectrometry)

Solid polyimide resins (5 g) were finely ground and pelletized, liquid samples were analyzed directly. K and Ca were measured by means of a *Siemens SRS-200* wavelength dispersive sequential spectrometer with chromium target tube (45 kV / 55 mA, LiF (100)-crystal 200 s under He).

RNAA (Radiochemical Neutron Activation Analysis

U and Th were determined after specific radiochemical separation of the indicator radionuclides ^{233}Pa and ^{239}Np for Th and U respectively[9].

3   RESULTS AND DISCUSSION

Determination of U and Th

Figure 2 shows count rates typically obtained in routine work for U and Th. 100 ng/l ^{233}U usually gives count rates between 400 and 800 counts/s. Figure 2a shows a calibration standard containing 250 ng/l U, Th and 100 ng/l ^{233}U. In an ashed sample shown in Figure 2b, Th-content was determined <25 ng/kg and U-content 105 ng/kg.

The method has been verified by recovery experiments and independent analyses performed by neutron activation analysis. Results of recovery experiments performed with 1 and 2 ng U and Th from elemental standards are shown in Table 2. Table 3 shows recovery of U (1 and 2 ng) added to polyimide sample. Results of both recovery experiments were satisfying. Results for U obtained by RNAA (Table 4) were slightly higher compared to ICP-MS results but still in satisfying agreement considering the fact that these contents are well below the ppb-range. Blank values for the described procedure are usually below 5 ng/kg for both radioisotopes. For determination of U and Th in polyimides, detection limits of 25 ng/kg for both elements can be guaranteed in routine analyses.

Table 2: Recovery of U and Th from standards via CPA/ICP-MS

element	addition / ng	measured / ng	recovery / %
Th	1	0.99	99
Th	2	1.93	96
U	1	1.01	101
U	2	1.99	100

a)

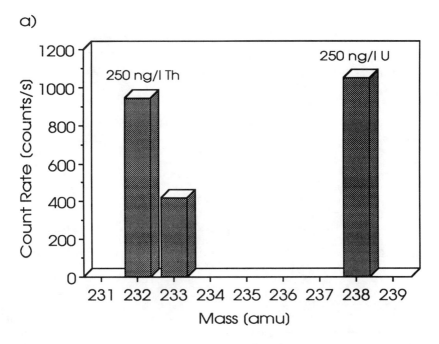

b)

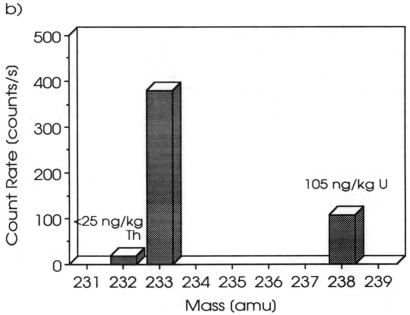

Fig. 2a: Calibration standard: 250 ng/l U, Th
+ 100 ng/l ^{233}U
2b: Ashed polyimide sample + 100 ng/l ^{233}U
(dilution factor: 4)

Table 3: Recovery of U added to polyimide sample (Th < 25
         ng/kg, U: 163 ng/kg). Sample weight was 2.07 g.

element	addition / ng	measured / ng	expected / ng	recovery / %
U	1	1.39	1.34	104
U	2	2.16	2.34	92

Table 4: Comparison of results for U in two polyimide sam-
         ples obtained by CPA/ICP-MS and RNAA, respectively.
         Th-contents were <25 ng/kg.

sample	U / ng g^{-1} ICP-MS	U / ng g^{-1} RNAA
1	0.30 ± 0.05	0.33 ± 0.05
2	0.16 ± 0.01	0.18 ± 0.01

Determination of Al, Ba, Ca, Cr, Cu, Fe, K, Li, Mg, Mn, Na, Ni, Pb, Sr, Ti and Zn

The procedure involving CPA/ICP-MS of "Determination
of U and Th" has been further optimized and evaluated in
order to investigate its capabilities for analyses of other
metal impurities of major interest (Al, Ba, Ca, Cu, Cr, Fe,
K, Na, Ni, Zn, Pb, Mg, Mn). Recoveries were studied with the

Table 5: Recoveries involving cool plasma ashing / ICP-MS
         Sample: SRM NIST 1634b *Trace Elements in Fuel Oil*

element	weight / ng	recovery / %
Na	9000	101
Al	1600	109
Ca	1500	133
Cr	70	114
Mn	23	87
Fe	3160	126
Co	32	91
Ni	2800	104
Zn	300	100
Pb	280	107

SRM NIST 1634b *Trace Elements in Fuel Oil* (Table 5) as well
as with multi-element standard solutions (Table 6).

Recoveries were within 10 % of the given values.
Higher recoveries for Fe and Ca in NIST 1634b (Table 5)
could be confirmed by XRFS and ICP-OES.   Results obtained
for Fe were 120 % (XRFS) and 114 % (ICP-OES) of the certified value.   Results for Ca were 140 % (XRFS) and 166 %
(ICP-OES) of the given content.

Table 6:   Element recovery involving CPA/ICP-MS

element	weight / ng	recovery / %
Na	2500	110
Mg	500	97.5
Al	2500	102
K	500	106
Ca	2500	113
Cr	500	107
Fe	500	99
Co	500	103
Ni	500	106
Cu	500	102
Zn	500	110
Cd	125	110
Mo	2500	91
Ba	2500	105
Pb	500	99

Another attempt to verify CPA/ICP-MS analysis
procedure was to compare Na, Fe and Cu results obtained by
ICP-MS with ZGF-AAS results (Table 7).   The results compare
quite satisfactorily.   The higher values listed do not
represent typical values for pure products. These samples
were analyzed with the purpose to investigate elemental
recoveries.   K and Ti results at levels of about 0.3 ppm
were confirmed by XRFS.

Detection limits of the metals analyzed are governed
by the capabilities of the ICP-MS instrument and by the
achievable blanks: keeping the blank values of the
ubiquitous elements as low as possible seems to be the most
crucial step to determine these elements at the lower ppb
level.   Due to the use of quartz vessels and a high sample
weight the blank values for Al, Ca, Cr, Fe, K, Na, Ni, Pb,
Ti and Zn could be kept ≤10 ppb and for Ba, Cu, Li, Mg, Mn
and Sr even <1 ppb (center column of Table 8).   The current
limits of determination (e.g. Ca, Cu and K; right column of
Table 8) are calculated on the basis of blank values and
their reproducibility during actual analytical runs.

Table 7: Determination of Na, Fe and Cu in several poly-
         imide samples. Comparison of results obtained by
         AAS and ICP-MS.

element	PTFE / AAS / ng g^{-1}	CPA / AAS / ng g^{-1}	CPA / ICP-MS / ng g^{-1}
Na	240	220	285
	230	220	305
	100		170
	510		600
	700		550
	500		500
	<100		<100
	<100		<100
Fe	470	385	400
	130		100
	400		400
	300		250
Cu	<50	10	11
	<50	17	18

Table 8 center column: Blank values for CPA/ICP-MS (blank
         values are set in relation to sample weight of 4 g)
         right column: Trace metal impurities determined in
         a polyimide production sample

element	blank / ng g^{-1}	content / ng g^{-1}
Al	5	60
Ba	<1	1
Ca	10[*)]	<100[*)]
Cr	6	70
Cu	<1	<10
Fe	10	100
K	8	<100
Li	<1	1
Mg	<1	35
Mn	<1	3
Na	5	170
Ni	2	15
Pb	4	<5
Sr	<1	<1
Ti	4	50
Zn	5	55

[*)] Ca was measured by ICP-OES after cool plasma ashing

Table 8 (right column) also shows results for metal impurities in a polyimide production sample with the sum of all analyzed metal impurities well below 1 ppm.

5 OUTLOOK

The perspective on this topic is that the requirements concerning the purity for these electronic chemicals will enhance further and that we have to continously improve the analytical procedures to keep up with the demands.

6 REFERENCES

1. V. Krivan, Nachr. Chem. Tech. Lab. 1991, 39, 536
2. OCG Electronic Chemicals Photoresists, Product Descriptions, OCG Microelectronic Materials Inc., CH-4002 Basle, Switzerland
3. Book of Semi Standards, 1987, 1, Chemicals Devision, Semiconductor Equipment and Material Institute, Inc., 805 East Middlefield Rd., Mountain View, CA 94043, USA
4. R. Iscoff, Semiconductor International, 1984, 7, 116
5. P. Burggraaf, Semiconductor International, 1988, 11, 58
6. H. Baumann, Chimia, 1990, 44, 236
7. H. Baumann, J. Pavel, Mikrochim. Acta [Vienna], 1989, III, 413
8. G. Knapp, Int. J. Environ.Anal. Chem., 1985, 22, 71
9. M. Franek, V. Krivan, to be published

ACKNOWLEDGEMENT

We wish to thank Prof. Dr. V. Krivan from the Sektion Analytik und Höchstreinigung, Universität Ulm, Germany, for RNAA analyses.

A Comparison of SCIEX Total Quant Results with Quantitative Results of Various Solutions by ICP-MS

A. V. Mistry and D. S. Lowe
DEFENCE RESEARCH AGENCY, ROYAL ARSENAL (EAST), METALS LABORATORY, WOOLWICH, LONDON SE18 6TD, UK

1   INTRODUCTION

Inductively Coupled Plasma Mass Spectrometry (ICP-MS) has become an established technique of multi-element trace analysis. It has been used for the analysis of a variety of sample types. These include organic materials[1], metals, ceramics and many others. Analysis of samples may be carried out by various modes, for example: Quantitative analysis, Isotope dilution, Isotope ratio and Total Quant.

The Sciex Total Quant program uses total mass spectrum for analysis. Isobaric and molecular interferences and the relative abundances of the isotopes are taken into account by sophisticated heuristics. The element intensities are assigned by initially obtaining the full mass spectrum, which results in a series of integrated intensities for each mass measured. A preliminary estimate is made for each element intensity based on isotopic natural abundances and the measured data. Initial estimates are made for polyatomic species. Intensity assignments are then made for multi-isotopic elements. Next, the data is examined for any changes necessary to the assignments for apparent inexact isotopic abundances. This is followed by assigning intensities for polyatomic and doubly-charged species with the multi-isotopic data. Monoisotopic elements and their associated doubly-charged and polyatomic ions are then assigned their intensities. To obtain the concentrations of the elements in a sample, a set of relative response factors for each element is applied to the element intensities to achieve the corresponding element concentration. These response factors can be calibrated using an external standard having a number of elements of known concentrations.
The aims of this work were as follows:

1.   To demonstrate the update of a 23 element based Total Quant program by five of these elements, namely Mg, Rh, Ni, Mo and Bi at 1 μg/ml level.

2.   To compare the results obtained by Quantitative Analysis with that of the Total Quant program, using high purity copper samples for trace elements and effluent water samples for mercury content.

2   EXPERIMENTAL

Instrumentation

The instrument used was the Perkin-Elmer ELAN 500 ICP-MS. This
is connected via a mass flow controller (Model 5876, Brooks
Instruments) to the nebuliser line and a peristaltic pump
(Gilson Manipuls 2) was set on the sample delivery tube to
provide a constant flow rate of 1.0 ml/min. The ICP-MS is
operated via the ELAN 5000 software. Optimisation was carried
out using a 0.1 ppm LCCP (Li, Cu, Cd, Pb and Rh) solution in
1% HNO_3. The ICP-MS instrumental parameters are shown in
TABLE 1.

Sample Preparation

A.   The 23 element Total Quant Calibration solution (1 μg/ml)
of the elements: Mg, Al, Ti, V, Cr, Mn, Fe, Co, Ni, Cu, Zn, As,
Se, Sr, Zr, Mo, Rh, Cd, Sn, Sb, Te, Pb, Bi, was prepared by
adding 100 μl of their respective 1000 μg/ml solutions to a
100 ml volumetric flask containing 5 mls of conc. HCl acid
(Baker) and made up to the mark. The five element 1 μg/ml
solution of Mg, Ni, Rh, Mo and Bi was prepared in a similar way.
Calibration check solutions (0.1 μg/ml) were prepared in a
similar manner.

B.   For the analysis of copper samples, the samples were first
pickled in dilute HNO_3 acid (Baker) to remove any contamination
from sample cutting. The flask used for making up the solutions
was cleaned by adding 20 mls of conc. HNO_3 acid (Baker) and
making up to the mark with water. The flask was allowed to
stand for approximately 20 minutes. The cleaned copper samples
were weighed accurately (0.500 g $\pm$ 0.005 g) into a beaker.
Dilute HNO_3 acid (10 mls of 50% v/v; Baker) was added and gently
warmed to aid dissolution. 1 ml of 10 μg/ml rhodium solution
was added to the cooled copper solution as an internal standard.
The solution was then diluted to 100 mls in a volumetric flask.
Sample dissolution for Certified Reference Materials was carried
out in the same way.

C.   The effluent water samples were prepared by adding dilute
HNO_3 acid (10 mls of 50% v/v; Baker) to 1 ml of the sample in a
beaker and heating the mixture to boiling, simmered for 20 mins
and finally cooled. 1 ml of 10 μg/ml rhodium solution was added
as an internal standard and the solution was made up to 100 mls.

Sample Analysis

A.   To demonstrate the update of the 23 element based Total
Quant program by a five element (Mg, Ni, Rh, Mo, Bi) solution at
1 μg/ml level, a calibration file containing the concentration
details of the five elements was set up. The five element
program was then used to determine the concentration of the
individual elements of the 23 element Total Quant calibration
solution (1 μg/ml) and the respective 0.1 μg/ml solution. The
results of the analysis are shown in TABLE 2. Measurement
parameters for the Total Quant Analysis are shown in TABLE 3.

B.   Quantitative Analyses of High Purity copper samples was
carried out using a conventional external calibration procedure.
The calibration standards (range 0-0.1 $\mu g/ml$) were prepared by
dilution of a 10 $\mu g/ml$ mixed standard containing Ni, As, Se, Sn,
Sb, Te, Pb, Bi. Rhodium was added as an internal standard at a
concentration level of 0.1 $\mu g/ml$. Calibration curves were
calculated from the intensity ratios of the internal Standard
and the analyte element. Certified Reference Materials from the
National Bureau of Standards were then analysed. These were
high purity coppers SRM C1251 and SRM C1252. The measurement
parameters are shown in TABLE 4. Total Quant Analysis of the
copper samples were carried out using the 23 element calibration
solution (1 $\mu g/ml$). Ultrapure water was used as a blank
solution.

A comparative set of Quantitative Analysis and Total Quant
results for the high purity copper samples are summarised in
TABLE 5.   SRM C1251 and SRM C1252 results are summarised in
TABLE 6.

C.   Quantitative Analysis of effluent waters for mercury
content was performed in the same way as that of the high purity
copper samples. External calibration standards were prepared in
the range (0-1 $\mu g/ml$). The measurement parameters for
Quantitative Analysis are shown in TABLE 7. A comparative set
of Quantitative Analysis and Total Quant results are summarised
in TABLE 8.

TABLE 1

ICP–MS OPERATING PARAMETERS

ICP operating conditions	
RF forward power	1250 W
Reflected power	< 5 W
Argon Flows	
Plasma gas flow rate	16 1/min
Auxiliary gas flow rate	1.4 1/min
Nebuliser gas flow rate	0.5 1/min
ICP–MS interface	
Sampling depth above load coil	10 mm
Sampling cone	Platinum with orifice of 1.14 mm
Skimmer cone	Platinum with orifice of 0.89 mm
Vacuum	
Mass spectrometer working pressure	4×10^{-6} Torr

TABLE 2
RESULTS OF THE 23 ELEMENT TOTAL QUANT
SOLUTION AT 0.1 μg/ml AND 1.0 μg/ml LEVEL

ELEMENT	[1] 0.1 μg/ml	[2] 0.1 μg/ml	[3] 1.0 μg/ml
Mg	0.09876	0.10050	0.96580
Al	0.09026	0.10840	0.86700
Ti	0.10180	0.10440	0.89560
V	0.19540	0.21420	0.86430
Cr	0.08780	0.08668	0.88850
Mn	0.10540	0.10120	1.00900
Fe	0.07663	0.07750	0.91150
Co	0.09005	0.08223	0.86510
Ni	0.09939	0.07780	0.98980
Cu	0.09399	0.08843	0.87860
Zn	0.13930	0.11150	1.24500
As	0.09499	0.11800	0.97010
Se	* ND	* ND	1.01000
Sr	0.10280	0.10430	1.01300
Zr	0.10670	0.10600	1.06100
Mo	0.10710	0.10480	1.06200
Rh	0.09864	0.10690	0.92790
Cd	0.11390	0.10570	1.03400
Sn	0.10570	0.10370	0.99950
Sb	0.10670	0.10040	1.05300
Te	0.10970	0.09535	1.14500
Pb	0.09591	0.10240	0.84490
Bi	0.09795	0.10330	0.86570

*Not detected
1.   Results obtained after calibrating with 5 element solution.
2.   Results obtained after calibrating with 23 element
solution.
3.   Results obtained after calibrating with 5 element solution.

TABLE 3
TOTAL QUANT ANALYSIS MEASUREMENT PARAMETERS

Scanning Mode	Normal
Replicate time	70 ms
Dwell time	7 ms
Sweeps/Reading	10
Readings/Replicate	1
No. of Replicates	2
Resolution	High (PW 0.6 amu at 10% PH)

TABLE 4
QUANTITATIVE ANALYSIS MEASUREMENT PARAMETERS
FOR HIGH PURITY COPPER SAMPLES

Scanning Mode	Normal Transient
Replicate time	900 ms
Dwell time	100 ms
Sweeps/Reading	3
Readings/Replicate	3
No. of Replicates	3
Resolution	High

TABLE 5
QUANTITATIVE ANALYSIS AND TOTAL QUANT RESULTS
FOR HIGH PURITY COPPER SAMPLES ($\mu g/g$)

	1	2	3	4	5	6	7	
^{58}Ni	3.2	2.1	2.9	1.8	1.7	10.8	9.9	Q
Ni	3.0	3.0	3.9	2.3	1.6	11.1	12.0	TQ
^{75}As	2.2	2.6	2.2	1.3	1.5	0.7	0.5	Q
As	3.9	3.5	3.0	1.5	2.2	0.4	0.7	TQ
^{78}Se	16.6	18.3	32.6	22.0	7.0	< 2	< 2	Q
Se	12.8	8.6	8.9	1.4	3.8	ND	ND	TQ
^{120}Sn	<1.0	<1.0	<1.0	<1.0	<1.0	<1.0	<1.0	Q
Sn	1.5	1.2	1.2	<1.0	<1.0	<1.0	<1.0	TQ
^{121}Sb	2.7	4.0	1.4	2.2	2.3	0.4	0.5	Q
Sb	4.4	4.6	1.3	2.2	2.1	0.4	0.6	TQ
^{128}Te	3.9	2.2	2.7	3.6	3.9	1.1	1.0	Q
Te	2.6	3.9	3.6	3.8	1.5	0.7	0.9	TQ
^{208}Pb	1.4	1.2	0.7	1.1	1.1	0.9	1.0	Q
Pb	1.2	0.7	0.4	0.7	0.6	1.4	1.1	TQ
^{209}Bi	<1.0	<1.0	<1.0	<1.0	<1.0	<1.0	<1.0	Q
Bi	0.3	0.2	0.2	0.3	0.3	0.2	0.3	TQ

1, 2, 3, 4, 5, 6, 7 Laboratory samples.   Q Quant Result
TQ Total Quant Result
ND Not detected

TABLE 6
QUANTITATIVE ANALYSIS AND TOTAL QUANT RESULTS
FOR SRM C1251 AND SRM C1252 ($\mu g/g$)

	SRM C1251				SRM C1252		
	^{1}Q	^{2}TQ	Certified		^{3}Q	^{4}TQ	Certified
Ni	9.68	9.34	22 + 2	Ni	64.8	61.6	128 + 3
As	9.20	9.29	(8̄)	As	75.3	54.8	124 ∓ 2
Se	2.00	ND	8.6+0.6	Se	18.1	ND	46 ∓ 2
Sn	8.20	8.03	(15̄)	Sn	70.5	85.4	(12̄4)
Sb	7.05	8.05	12.6+0.7	Sb	22.6	29.7	42 + 3
Te	10.2	7.00	(12̄)	Te	32.45	22.9	(4̄4)
Pb	5.58	7.12	7.5+1.5	Pb	40.6	56.0	60 + 2
Bi	3.70	2.69	(3̄)	Bi	15.9	18.0	20 ∓ 4

(   ) Nominal values

1, 2, 3, 4 Average figures of three determinations

TABLE 7
QUANTITATIVE ANALYSIS MEASUREMENT PARAMETERS
FOR EFFLUENT WATER SAMPLES

Scanning Mode	Normal Transient
Replicate time	1800 ms
Dwell time	200 ms
Sweeps/Reading	3
Readings/Replicate	3
No. of Replicates	3
Resolution	High

TABLE 8
ANALYSIS OF EFFLUENT WATERS FOR MERCURY CONTENT
QUANTITATIVE ANALYSIS AND TOTAL QUANT RESULTS ($\mu g/ml$)

	^1Q	TQ
Sample A	< 0.02	0.020
Sample B	< 0.02	0.010
Sample C	< 0.02	0.010
Sample D	< 0.02	0.010
Sample E	< 0.02	<0.005
Sample F	< 0.02	0.031
Sample G	< 0.02	0.014
Sample H	< 0.02	0.011
Sample I	< 0.02	0.013
0.1 $\mu g/ml$ Hg Standard	0.0981	0.0989

1. ^{202}Hg isotope

3   RESULTS AND DISCUSSION

A.   The results of the 23 element Total Quant solutions at
0.1 $\mu g/ml$ and 1.0 $\mu g/ml$ levels (TABLE 2) show that the 5 element
standard (1.0 $\mu g/ml$) can be used to correctly update the entire
response table.   This observation is also indicated by the good
comparison of the at 0.1 $\mu g/ml$ level based on 23 element and
5 element (Mg, Ni, Rh, Mo, Bi) calibration solutions, with the
exception of Vanadium.   The most abundant isotope (97.76%) $^{51}V^+$
suffers from $^{35}Cl^{16}O^+$ interference; the second most abundant isotope
(0.24%) $^{50}V^+$ overlaps with $^{35}Cl^{15}N^+$ as well as $^{50}Ti^+$ and $^{50}Cr^+$.
This may be the reason for the relative high vanadium content at
0.1 $\mu g/ml$ level, suggesting the Cl content needs to be matched
closely for samples and standards.

B.   The analysis of high purity coppers (TABLES 5 and 6)
summarises a good agreement between Total Quant and Quantitative
results.   The variation of Se content may be explained by the
possible $^{40}Ar^{36}H_2^+$ interference on $^{78}Se^+$, $^{40}Ar^{40}Ar^+$ on $^{80}Se^+$, ie
in the absence of an interference free Se isotope, the accuracy
is impaired.

C.   The effluent waters analysis for mercury content (TABLE 8)
shows a comparative set of Total Quant and Quantitative results,
indicating the usefulness of the two options for mercury
determinations.

4   CONCLUSION

Experience shows that when analytes are determined in a sample
by ICP–MS, the Quantitative Analysis option demands the
specification of the isotopes to be measured, the content of the
standards, the elemental equations required to correct for
isobaric interferences. The Total Quant program, however, is
unique. It analyses the whole mass spectrum and quantifies all
the elements. This option also calibrates for all elements
using concentration data based on a few. This comparison of the
two options has been demonstrated by the work presented.

The Total Quant program can be quite useful for the full
analysis of samples, saving valuable time for analytical method
development.

The comparative set of results obtained by analysing the high
purity copper samples and the effluent waters indicate the
possibility of analysing, other matrix types successfully by
either of the options.

5   REFERENCES

1.   D.S. Lowe and R.G. Stahl, Anal. Proc., 1992, 29, 277.

A Comparison of ICP-MS with Neutron Activation Analysis for Multielement Determination in Groundwaters

T. Probst, P. Zeh, and J. I. Kim
INSTITUT FÜR RADIOCHEMIE, TECHNISCHE UNIVERSITÄT MÜNCHEN, WALTHER-MIEßNER-STR. 3, D-8046 GARCHING, FEDERAL REPUBLIC OF GERMANY

1   Introduction

A recent development of inductively coupled plasma mass spectrometry (ICP-MS) has made its application much easier, with its high sensitivity for many elements, for the multielement analysis of inorganic trace constituents in neutral aquatic samples. The sensitivity of ICP-MS approaches nearly that of reactor neutron activation analysis (NAA) for a large number of elements but is still inferior to the latter in one or more orders of magnitude in general. The main advantage and disadvantage of the two methods are obvious: ICP-MS is fast in analysis, while NAA is contamination free in operation. The accuracy of ICP-MS analysis for its sensitivity range cannot be, for obvious reasons, easily verified by other laboratory means but NAA for the moment.

The comparison of multielement analysis by ICP-MS as well as by NAA can be found in the intercomparison exercise for the IAEA reference sample human diet H9 [1] and for biological reference samples [2]. The two methods are compared for the analysis of rare earth elements in bauxites [3] and for some elements in thermal waters [4].

The present paper concentrates on the analysis of deep groundwaters rich in aquatic colloids of organic nature (humic substances), in which the main part of heavy trace elements are found as a colloid-borne chemical state. Both ICP-MS and NAA are applied for the quantification of inorganic trace elements in various ultrafiltration fractions. Thus the comparison of analytical results follows the different concentration ranges from µg/L to ng/L.

2   Experimental

The sampled aquifer from borehole Gohy 2227 in the Gorleben region, northern Germany, intersect a relatively deep and saline water system with glacial intercalations of lignite. From borehole Gohy 2227 the water was collected at depth below the surface of 130.5 m - 131.2 m. The groundwater Gohy 2227/3 rich in humic substances and in salinity is used for a comparison of ICP-MS with Monostandard-NAA (MS-NAA) for multielement determination. The groundwater is submitted to anaerobically ultrafiltration from pore size of 1000 nm to 1 nm to examine the concentration change in colloid-borne elements in filtrates. MS-NAA is performed by irradiating the groundwater filtration series and gamma spectrometric assay of the sample without preconcentration or chemical separation. A well proven monostandard method is used for the quantification of each element.

ICP-MS:

The ICP-MS mass spectrometer of Perkin Elmer & Sciex Elan 5000 was used in its standard configuration with a cross-flow nebuliser and Pt-cones. The operating conditions are summarized in Table 1. Multi-element standards were prepared in 2 % v/v nitric acid by dilution of 1000 µg/mL standard solutions for atomic absorption spectroscopy (Aldrich) of individual elements. Blank, standard and sample solutions were spiked with 10 µg/L Rh as an internal standard. This was a solution of $(NH_4)_3$-$RhCl_6$ in 5 % v/v hydrochloric acid. All groundwater samples were analysed using external calibration as well as standard addition calibration. Sample treatment and ultrafiltration are shown in Figure 1.

Table 1: ICP-MS operating conditions

ICP-System:			Mass spectrometer:		
sample uptake	[mL/min]	1.1	CEM	[kV]	4.65
plasma power	[W]	991	Deflector	[kV]	2.371
Ar flows:			Running vacuum	[Torr]	$6.66 \cdot 10^{-6}$
nebuliser	[L/min]	0.890	Basic vacuum	[Torr]	$4.67 \cdot 10^{-7}$
auxillary	[L/min]	0.9	Ion lens voltages:		
plasma	[L/min]	14	B, P, E1, S2	[V]	45, 43, 23, 45
Peak scan parameters:					
replicates		3			
dwell time	[ms]	50			
analysis time	[min]	8.32			

MS-NAA:

The method of MS-NAA [5] used in the present experiment consists of irradiating a large volume of water samples (250 mL), measuring by multi-nuclide gamma spectrometry and evaluating concentrations by making use of relevant nuclear data [6]. The whole procedure is non-destructive. Ultrafiltrates are sampled in a 250 mL "Suprasil" quartz (Heraeus Co., Hanau, FRG) irradiation flask. The water filled flask is covered with a "Suprasil" quartz cylindrical cap and then irradiated in the Forschungsreaktor München. As a monostandard and a flux monitor, the known amounts of Au-Al alloy (0.1096 ± 0.0004 % Au) and Co-Al alloy (0.1023 ± 0.0008 % Co) wires of 0.5 mm diameter are put into a quartz capillary and placed in the center of the water sample. The whole irradiation and detection systems are described in detail elsewhere [5 -7].

3 Results and Discussion:

The limits of detection listed in Table 2 are demonstrating the analytic capabillity of ICP-MS compared with MS-NAA [9]. The sensivity of Ni, Nd and Ta approaches, even in the two groundwater samples, nearly that of MS-NAA. The limits of detection for samples with simple matrix shown in column four are obtained by using a different nebulizer type. Peak form and peak noise are analysed in detail with the scanning mode.

Table 2: Limits of detection for ICP-MS data compared to MS-NAA

	Sample 573[1] ng/L	Sample 2227[1] ng/L	in simple matrix[2] ng/L	Literature values ng/L	NAA[3] ng/L
Sc	500	n. d.	100	20-100	0.004
Cr	$10*10^3$	n. d.	130	20	0.6
Mn	n. a.	n. a.	40	2-40	13
Fe	$8*10^3$	$10*10^3$	60	200-400	40
Co	200	300	10	0.9-20	0.1
Ni	60	50	40	5-30	2
Zn	$2*10^3$	$3*10^3$	50	3-80	5
Rb	10	20	30	3-20	0.1
Sr	$10*10^3$	$20*10^3$	120	0.8-20	53
Zr	$6*10^3$	$8*10^3$	610	4-30	30
Sb	60	50	60	1-20	0.02
Cs	n. d.	10	10	0.5-10	0.09
Ba	$2*10^3$	$5*10^3$	470	2-20	30
La	60	20	30	0.5-20	0.01
Nd	10	10	20	2-20	3
Eu	80	40	2	0.7-20	0.03
Tb	30	70	30	0.5-10	0.09
Yb	40	60	10	1-20	0.04
Lu	30	50	4	0.5-10	0.03
Hf	20	80	6	0.6-30	0.09
Ta	20	40	40	0.6-20	0.2
Th	200	300	20	0.5-20	0.07
U	100	70	30	0.5-20	0.06

[1]: cross-flow nebulizer
[2]: Meinhard nebulizer; scanning mode
[3]: NAA-detection limits for pure water analysis [7]

Our initial investigations of ICP-MS deals with different standard reference materials. Precision of 4 % or better was achieved, after optimizing the ICP-MS, with the NBS SRM 1643C water reference standard, with the aqua regia solute of BCR CRM 141 and BCR CRM 142. The ICP-MS parameters shown in Table 1 are obtained by external calibration and standard addition calibration for determining the reference water standard NBS 1643C. These parameters are also used to determine the element concentration of the groundwater filtrates.

The applicability of ICP-MS is assessed by comparing ICP-MS results of uncharacterized samples with those obtained from MS-NAA. To avoid systematic errors resulting from different sample pretreatments (digestion, fusion), liquid samples are investigated independently by ICP-MS and MS-NAA. The analytical capability of ICP-MS is appraised for more than 20 elements in the concentration range of µg/L to ng/L in particular groundwaters. The groundwater 2227/3, rich in humic substance and salinity, is selected to examine relatively high matrix effects of organic and inorganic nature. The detection limits of ICP-MS are compared to those of MS-NAA. It is known from earlier work [8] that trace elements of higher oxidation states are, in general, preferentially associated with humic colloids. Ultrafiltration with different pore sizes lower the humic colloid contents in the water samples together with the concentrations of polyvalent cations, because these ions are colloid-bounded.

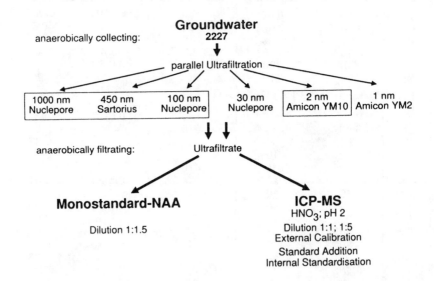

Figure 1: Sample treatment and ultrafiltration

Table 3: Groundwater 2227/3, element concentrations in µg/L

	NAA	ICP-MS	NAA	ICP-MS
	1000 - 100 nm Filtrates		2 nm Filtrate	
Fe	283 ± 9	308 ± 8	41 ± 1	40 ± 10
Co	0.58 ± 0.08	0.61 ± 0.03	0.282 ± 0.003	0.38 ± 0.08
Zn	55 ± 7	48.4 ± 0.5	13 ± 8	12 ± 1
Rb	2.5 ± 0.3	2.29 ± 0.03	2.5 ± 0.5	2.27 ± 0.09
Sr	260 ± 10	266.5 ± 0.5	215 ± 20	238 ± 1
Zr	90 ± 2	101.1 ± 0.6	14 ± 2	25 ± 1
Sb	0.06 ± 0.03	0.06 ± 0.06	0.03 ± 0.01	n. d.
Cs	0.06 ± 0.04	0.06 ± 0.02	0.06 ± 0.04	0.07 ± 0.05
Ba	32.3 ± 0.6	41.2 ± 0.4	30 ± 3	34 ± 1
La	2.35 ± 0.06	2.52 ± 0.04	n. d.	0.38 ± 0.09
Nd	3.1 ± 0.2	2.77 ± 0.03	n. d.	0.45 ± 0.08
Eu	0.27 ± 0.03	0.30 ± 0.04	0.035 ± 0.001	0.02 ± 0.03
Tb	0.23 ± 0.01	0.24 ± 0.03	0.023 ± 0.003	n. d.
Yb	1.08 ± 0.04	1.23 ± 0.04	0.171 ± 0.002	0.3 ± 0.3
Lu	0.333 ± 0.009	0.22 ± 0.04	0.054 ± 0.001	0.06 ± 0.01
Hf	1.01 ± 0.02	1.31 ± 0.06	0.205 ± 0.003	0.2 ± 0.1
Ta	0.039 ± 0.002	0.05 ± 0.06	n. d.	n. d.
Th	1.76 ± 0.08	1.63 ± 0.09	0.038 ± 0.001	0.09 ± 0.09
U	1.0 ± 0.2	0.85 ± 0.07	n. d.	0.07 ± 0.07
DOC Σ Σ	79.9 ± 0.3 mg/L Cations 44.3 meq/L Anions 44.9 meq/L		40.9 ± 1.1 mg/L	

Table 2 summarizes the results from MS-NAA and ICP-MS analysis of two ultrafiltrates of the water sample 2227/3. The elements are listed in the mass order.

The values summarized in the NAA column for the 100 nm filtrate in Table 2 are the mean values of the independently determined results for 1000 nm, 450 nm and 100 nm filtrates. The ICP-MS values of this filtrate are analysed in fourfold. An unfiltered sample is investigated in addition to the filtration series (> 100 nm). No obvious concentration change could be detected with both analytical techniques in the ultrafiltrates from > 100 nm pore size.

Within the column of the 2 nm filtrate the concentrations of La, Nd, and U are only determined by ICP-MS. For Cs, Eu, Yb, Hf, Th, and U the ICP-MS standard deviations are very close to the analysed concentration values. La, Nd couldn´t be detected by MS-NAA because of the chosen irradiation time which affects a cooling time greater than half lives of these activated nuclides.

Figure 2 compares the element concentrations obtained from ICP-MS and MS-NAA with one another on a log-log scale. The results of MS-NAA are plotted on the x-axis. In general, the intercomparision of the results appears to be satisfactory for the element concentrations in the µg/L region and below. In the filtrate from 2 nm pore size, the total number of analysed elements decreases but the regression line with slope 1 and intercept 0 is still well fitted for most elements.

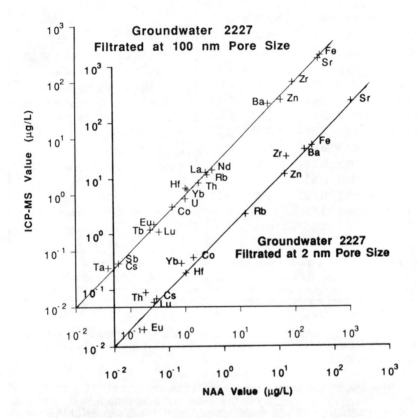

Figure 2: ICP-MS values vs MS-NAA for the sample 2227/3

Ultrafiltration doesn't affect the monovalent Rb and Cs concentrations very much. The concentrations of the divalent cations, Sr and Ba, are lowered insignificantly. The analysed element values obtained by ICP-MS and MS-NAA are controlled additionally by this behaviour [8].

A more detailed picture is given in Figures 3 and 4, which present a comparison of the two corresponding analytical values for each element by the following relations:

$$RD\ (\%) = \frac{value_{ICP\text{-}MS} - value_{NAA}}{value_{NAA}} * 100$$

$$S_{RD} = RD * \sqrt{\frac{s^2_{ICP\text{-}MS} + s^2_{NAA}}{(value_{ICP\text{-}MS} - value_{NAA})^2} + \frac{s^2_{NAA}}{value^2_{NAA}}}$$

In the sample 2227/3 (Figure 3) ultrafiltrated at 100 nm pore size, almost all element concentrations are within 20 % standard deviation (S_{RD}). Only for Ta the S_{RD} value is significantly higher. The overall tolerance of ICP-MS values differing from MS-NAA are acceptable in this particular experiment within ± 20 % relative deviation (RD).

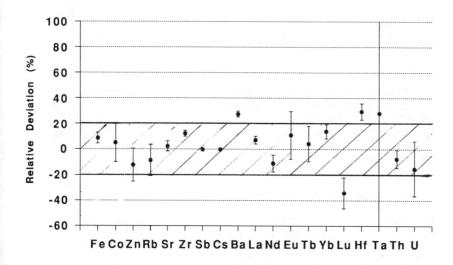

Figure 3: Groundwater 2227/3 filtrated at 100 nm pore size

The ultrafiltrate from 2 nm pore size (Figure 4) shows higher values of RD and S_{RD} because of low element concentrations being close to the limits of determination of ICP-MS for many elements. For a large number of elements the 20 % range of RD is not altered but the S_{RD} value becomes considerably larger. Within this ng/L concentration range, the ICP-MS determination is closely limited for many elements.

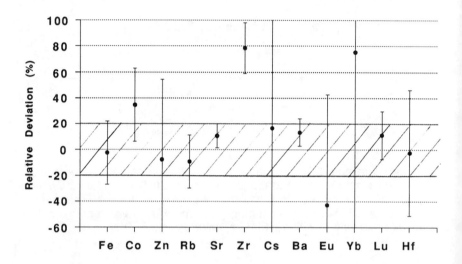

Figure 4: Groundwater 2227/3 filtrated at 2 nm pore size

Standardization in the ng/L area, raising the number of internal standards, using ultrasonic nebulization, microsampling by flow injection techniques, and working with multivariate calibration methods (GSAM and PGSAM) [10], gives way to analyse ng/L concentrations with ICP-MS even with samples of difficult matrices. We have incorporated these methods into our studies.

4   Acknowledgements

We would like to acknowledge cand. dipl. chem. M. Rupprecht for multivariate calibration method development and Dr. K. Cervinsky for manuscript correction.

This work was financially supported by the Commission of the European Communities´ R&D Programs on Management and Storage of Radioactive Waste.

5   Literature

[1]   J. J. Fardy, I. M. Warner, J. Radioanal. Nucl. Chem. Art., 1992, 157, 239.
[2]   N. I. Ward, F. R. Abou-Shakra, S. F. Durrant, Biol. Trace Elem. Res., 1990, 26, 177.

[3]   M. Ochsenkühn-Petropoulou, K. Ochsenkühn, J. Luck, Spectro-
        chim. Acta 1991, B46, 51.
[4]   E. Veldeman, L. Van't dack, R. Gijbels, M. Campbell, F. Van-
        haecke, H. Vanhoe, C. Vandecasteele: "Analysis of Thermal
        Waters by ICP-MS" in Applications of Plasma Source Mass
        Spectrometry ed. G. Holland, A. N. Eaton, Royal Soc. Chem.
        Bookcraft (Bath) Ltd., Cambridge, 1991, p 25.
[5]   J. I. Kim, H. Stärk, I. Fiedler, Nucl. Instr. Methods, 1980, 177, 557.
[6]   J. I. Kim, H. Stärk, I. Fiedler, H.-J. Born, D. Lux, J. Environ. Anal.
        Chem., 1981, 10, 135.
[7]   J. I. Kim, G. Böhm, R. Henkelmann, Fresenius Z. Anal. Chem.,
        1987, 327, 495.
[8]   J. P. L. Dearlove, G. Longworth, M. Ivanovich, J. I. Kim, B. Dela-
        kowitz, P. Zeh, Radiochim. Acta, 1991, 52/53, 83.
[9]   G. Böhm, J. I. Kim, J. Kemmer, Nucl. Instr. Methods Phys. Res.,
        1991, A305, 587.
[10]   M. E. Ketterer, J.J. Reschel, Anal. Chem., 1989, 61, 2031.

ICP-MS Certification of Toxic Metals on Filters

P. J. Paulsen and E. S. Beary
NATIONAL INSTITUTE OF STANDARDS AND TECHNOLOGY, INORGANIC
ANALYTICAL RESEARCH DIVISION, GAITHERSBURG, MARYLAND
20899, USA

1 INTRODUCTION

Inductively coupled plasma-mass spectrometry (ICP-MS) was used to determine the amounts of four trace elements; Pb, Cd, Zn and Mn on filters. The analytical potential of ICP-MS has been recognized by researchers since the early 1980's[1]. It has become a powerful tool in inorganic analyses since its commercial introduction in 1983. The ease of sample introduction at atmospheric pressure, high sensitivity and multi-elemental coverage of ICP-MS offer distinct advantages over traditional mass spectrometry for many analytical applications. However, difficulties with the stability of signal transmission of the ICP ion source requires a well defined analytical scheme to obtain high accuracy quantitation. The work presented here will describe the analytical procedures used to meet the stringent accuracy requirements for the certification of Pb, Cd, Zn and Mn in a renewal Standard Reference Material (SRM) 2976d, Toxic Metals on Filters.

Description and Preparation of SRM 2676d. This SRM is an analytical standard used in the validation of potentially hazardous metals in industrial atmospheres. Each SRM consists of a set of six membrane filters (two of each of three levels) on which known amounts of the elements of interest have been added. In addition, two blank filters are supplied with each set. The membrane filters are composed of biologically inert mixtures of esters of cellulose acetate and nitrate. The SRM preparation procedure will be described briefly, since it was used to address and assess the accuracy of the measurement system.

More than 8,000 filters were purchased from Millipore, Inc.[*] The 37 mm diameter filter has a pore size of 0.80 μm. A single lot was specified to prevent any variability in trace contamination due to the manufacturing processes. Two thousand one hundred filters were prepared for each of the three different levels of the four elements of interest. Three separate master solutions, one for each level, were gravimetrically prepared from high purity metals of Pb, Cd, Zn and Mn of known assay and purity. The final solution concentrations were prepared so that approximately 0.05 g of solution would deliver the desired amount of each element. Each filter was prepared by depositing a fixed volume of the appropriate master solution. The three doped levels were labelled Levels I-III. All added elements were in μg quantities (from 1 μg Cd in the lowest level to 100 μg of Zn in Level III). In addition, 2100 blank filters were prepared by depositing 0.05 g of 2% HNO_3 onto each filter. Approximately ten percent of the pipetted volumes were weighed so that the amount of each element in μg could be calculated. The range of 216 weights during the preparation of the 2100 filters for each level was about 0.0010 μg, or approximately 2% of the 0.05 g target. Most of the observed weights were toward the low end of the reported range. The average of the actual weights at each level varied slightly, presumably due to differences in the density of each master mix.

2 EXPERIMENTAL

<u>Materials and Reagents</u>. All of the mineral acids and high purity water used in this work were prepared and analyzed at NIST[2]. All experiments were performed in a Class 10 Clean Laboratory[3].

<u>Instrumentation</u>. The ICP-MS used for this work was PQ II Turbo Plus Mass Spectrometer from FISONS Instruments. The operating parameters were an rf power of 1.3 kW and a coolant, auxiliary, and nebulizer argon flow of 16, 0.5 and 0.85 L/min, respectively. A peristaltic pump fitted with polyvinylchloride (PVC) tubing provided a constant sample flow rate of 0.3 mL/min into a concentric nebulizer and a water cooled spray chamber at 2°C. The mass calibration was adjusted across the desired mass scale using a solution standard containing a selected group of elements in the mass region of interest. All data were collected in the peak jump mode. In a well calibrated system, the peak jump mode provides higher sensitivity as well as better accuracy and precision than the scan mode over an extended mass region (more than one element).

---

[*] Certain commercial equipment, instruments, or materials are identified in this report to specify adequately the experimental procedure. Such identification does not imply recommendation or endorsement by the National Institute of Standards and Technology, nor does it imply that the materials or equipment identified are necessarily the best available for this purpose.

Internal Standards Quantitation. Two different methods of quantitation were used during this study: an internal standard technique and isotope dilution. Manganese, mononuclidic at mass 55, was quantified versus [59]Co as an internal standard. ICP-MS quantitation based on signal intensity (eg. standard addition procedures or the use of external standards) depends upon the stability of the absolute signal intensity. This approach to quantitation is problematic in the ICP-MS because the signal transmission drifts with time. The addition of an internal standard minimizes the effects of this drift since the relative transmission of the standard versus the analyte normalizes the quantitative data. However, the stability of the relative transmission must be verified.

Isotope Dilution Quantitation. For multi-isotope analytes, isotope dilution is always the method of choice when high accuracy quantitation is required. In this work, Pb, Cd and Zn were quantified by isotope dilution. In isotope dilution, an enriched isotope of the analyte serves as an ideal internal standard since both the standard and the analyte have exactly the same chemical properties, exhibiting identical chemical behavior from the sample preparation to instrumental analysis. Isotope dilution is immune to any instrumental or sample variations that cause shifts in signal levels since an isotopic ratio of the element is used to quantify the analyte and not an absolute signal intensity[4]. Isotope dilution is the only quantitative technique which takes full advantage of the ICP mass spectrometer. Under carefully controlled conditions the ratio measurements are precise to about 0.2%.

Sample Preparation. A single spike mix was prepared containing [67]Zn, [111]Cd, and [206]Pb as well as [59]Co, in proportions suitable for spiking all three levels of filter samples. The blank was spiked with a quantitatively prepared dilution of this master mix. The filter samples were dissolved using high purity $HNO_3/HClO_4$ reagents. Each digested sample was evaporated to dryness and the residue redissolved in 2% HNO_3, so that the [67]Zn concentration was 20 ppb in solutions for the Zn and Pb determinations. The Cd and Mn were determined from solutions three times more concentrated. The minimal sample manipulation required for this work greatly decreased the risk of errors which can arise from complicated chemical separations and/or handling during sample preparation.

Calibration of Enriched Isotopes. The spike mix was calibrated against two independently prepared solutions containing the respective natural metals of known source and isotopic composition (particularly significant for Pb). The concentrations of [67]Zn, [111]Cd and [206]Pb enriched isotopes in the spike mix were determined versus the gravimetrically prepared natural solutions using isotope dilution analysis. The spike calibration usually reflects the ultimate measurement precision for an analyte under ideal conditions. Matrix affects on ion transmission as well as inhomogeneity of analytes in "real" samples can degrade measurement precision.

RSF Calibration of Mn versus Co. The Co in the spike mix was quantitatively prepared, and was also used in the calibration of the relative sensitivity factor (RSF). The RSF of ^{55}Mn/^{59}Co was calibrated using mixes of gravimetrically prepared solutions of each element. Two different solutions of each of the two elements were prepared from metals of known assay and chemical purity. These four solutions were used to gravimetrically prepare a total of eight mixes of the two elements. The relative sensitivity factor was determined based on the calculated Mn/Co ratio and the experimental or measured Mn/Co ratio, with precision of 0.2% rsd (the best precision expected for spike calibrations). There were no distinguishable differences between each pair of prepared natural standards. This experimentally determined RSF then served to correct for any shifts in the relative transmission of the analyte versus the internal standard during the analysis. Since the Co was present in the spike mix, and Mn was in the natural solution used as the calibrant, a spike calibration mix served as an isotopic standard to correct for mass discrimination (and shifts in mass discrimination) for all four analytes throughout the analysis.

3 RESULTS AND DISCUSSION

Instrumental Measurement Precision. The ICP-MS measurement precision for all four elements and each of the three doped levels is shown in Table I, as well as the uncertainties on the spike calibrations. As a check on the accuracy of measured values, the actual amount of each element delivered to the filters could be calculated using the concentrations of the master solutions and the nominal weight of solution delivered to each filter. The precision of the 126 pipetted weights were about 0.5% for each level and are listed separately in the Table. If the spike calibration is indicative of the measurement capabilities then the predominant source of uncertainty in the three filters levels can be attributed to the pipetting procedure.

Table I
ICP-MS Measurement Precision
(% rsd)

	Zn	Cd	Pb	Mn	Pipet Precision
Level I	0.59%	0.56%	0.44%	1.3%	0.46%
Level II	0.60%	0.55%	0.60%	0.72%	0.48%
Level III	0.27%	0.46%	0.42%	0.39%	0.51%
Calibrations	0.36%	0.14%	0.31%	0.20%	---

Individual results and uncertainties for the Zn determinations in all four levels (three doped levels and the blank level) are shown in Table II. Zinc was of particular interest since $ZnCl_2$ is used as a catalyst in the manufacturing process and was expected to be at measurable levels in the filter material. Individual results from the ten randomly selected "blank" filters (no added Zn in Level IV) show some evidence of inhomogeneity. However, even the 11% uncertainty in the blank measurement adds < 0.3% uncertainty to the lowest level filter, Level I, which does not jeopardize the usefulness of this SRM. Results of the ^{67}Zn calibration, using two separate natural solutions as previously described, are listed in the last column as an indication of the best precision expected under the analytical conditions selected.

Table II
Zinc by ICP-MS Isotope Dilution (µg)

	Level I	Level II	Level III	Filter Blank	Spike Calibration
	10.271	49.403	99.795	0.315	1.600312
	10.188	49.305	99.630	0.261	1.605145
	10.102	49.864	100.200	0.229	
	10.140	49.891	99.895	0.253	1.599649
	10.151	49.164	99.709	0.266	1.591168
	10.141	49.182	99.699	0.233	
	10.093	49.335	99.460	0.238	
	10.074	49.060	99.171	0.247	
	10.099	49.749	99.841	0.277	
	10.078	49.370	99.827	0.304	
Avg	10.134	49.432	99.723	0.262	1.599069
s	0.060	0.298	0.273	0.029	0.005809
rsd	0.59%	0.60%	0.27%	11.0%	0.36%

Table III lists all the analytical results obtained for all elements in Level III. As expected, the measurement precisions are among the best for this highest doped level, since predominant sources of imprecision remained constant for all levels. The reported Mn content in this level was of particular interest. The bracketed value 19.427 µg Mn was determined in a sample that was charred during an evaporation step. Subsequent redissolution of the sample was visually incomplete. Differences in chemical behavior between the analyte (Mn) and the internal standard (Co) prevented accurate quantitation of Mn in this sample. As expected, the IDMS values were not affected. Inclusion of this inaccurate value in the final results for Mn would have biassed the average by -0.2%. This exercise demonstrates that the accuracy of isotope dilution quantitation is independent of analyte recovery after isotopic equilibration.

Table III
Level III: Toxic Metals on Filters (µg)

	Mn (µg)	Zn (µg)	Cd (µg)	Pb (µg)
	20.088	99.795	10.092	29.887
	19.892	99.630	10.075	29.812
	19.917	100.200	10.177	30.114
	20.061	99.895	10.127	29.977
	19.927	99.709	10.094	29.869
	(19.427)**	99.699	10.087	30.188
	19.844	99.460	10.040	29.850
	19.874	99.171	10.062	29.924
	19.944	99.841	10.185	30.071
	19.926	99.827	10.124	30.021
Avg =	19.941	99.723	10.106	29.971
s =	0.08	0.273	0.047	0.124
rsd =	0.39%	0.27%	0.46%	0.42%

Analytical Blanks. There were two types of blanks considered in this analysis. One was the filter blank (Level IV of the SRM) and the other was the analytical blank due to reagents and sample processing. The average analytical blank for Pb, Cd, Zn and Mn was 0.01%, <0.04%, 0.09% and 0.02% respectively of the lowest filter level, Level I. For all elements, the analytical blank was subtracted from the reported value. All blanks were spiked at a level 100 times lower than the lowest level filter. Using this method for blank determinations, high precision results are neither expected nor required. The measured ratios are often very close to the spike ratio if the blank is significantly less than the level spiked for (100 times lower than the lowest level). Blanks which are high relative to the associated sample are measured with highest precision since the ratio is necessarily within a measurable range. Under some conditions, blanks are deliberately overspiked (which degrades precision) in order to obtain adequate spike signal intensity for measurement.

Filter Blanks. It is clear from the analytical results provided for the analytical blank and Level IV in Table IV, that Pb, Mn and Zn are present in the filters themselves at the levels indicated. The Cd value is indistinguishable from the

---

[**] Incomplete recovery of this sample prevented accurate quantitation of Mn. The IDMS values are not affected.

analytical blank. The filter blanks (Level IV) are an insignificant source of contamination for the Mn, Pb, and Cd in the three doped levels. However, Zn is present at high levels (about 0.3 μg) in the filter paper, and must be considered when the analytical results are compared with the results based only on the added amount.

Table IV
Blanks (μg)

	Mn	Zn	Cd**	Pb
Unprocessed Filters, BI 1-3, **Batch A**				
Average	0.0026*	0.626*	≤0.0004	0.0016*
rsd	4.4%	59%	≤0.0004	33%
Reagent (analysis) blank, BI 4-7				
Average	0.0005	0.009	≤0.0004	0.00075
rsd	17%	118%	≤0.0004	100%
Level IV, 8-18, **Batch B**				
Average	0.0048*	0.262*	-----	0.0024*
rsd	15%	11%	≤0.0004	47%

Full Mass Scans of "blank" filters. An additional study was performed to further assess the filter blank, since the Zn content was variable and significant. Two different 37 mm, type AA millipore filters were analyzed, labelled batch A and batch B. The two filters were exactly the same type, but not of the same batch. Filter B was from the lot presumably used for production of this SRM. However, during the course of the analysis the question arose as to whether or not the Level III filter was from the batch A or batch B. Filter A contained about 0.6 μg of Zn, while the Zn content in filter B was 0.3 μg as determined by IDMS. Since $ZnCl_2$ is used as a catalyst in the manufacturing process of these filters, batch to batch variation below 1 μg would be expected. A full mass scan was run on filters A and B under exactly the same conditions to provide relative concentrations without absolute quantitation. Immediate analysis of the Level III filter under the same conditions provided a means of identifying the filter source (batch A or batch B). The Mn and Pb levels were about the same in both filters A and B, and the Cd concentration indistinguishable from the background (see Table IV). Sodium, Mg and Al were detected at the same level in both filters at one to two orders of magnitude above background. Ultratrace and trace amounts of Fe, Ni, Sr, Zr, Mo, Ba, La and W were detected at about the same level in both filters at about ten times above background. The stainless steel drum and tungsten roller used in processing are the probable sources of this trace metal contamination. The Cu in Filter A was about six times higher than the Cu detected in Filter B. Using this comparative scan technique relative amounts of contaminants can be determined without quantitation. In comparing the three mass scans in Figure 1, it is evident that the Cu content in scan C (Level III filter) more closely matches the Cu content in scan B (Filter B). Scan A (Filter A) has a significantly higher Cu content. Using this simple ICP-MS procedure we were able to verify that the Level III filters were from batch B, the same lot as the other filters used for the SRM.

Comparison of Added/Measured Values. The ratios of the added versus the measured amount were calculated. Since the filters were acid digested for instrumental analysis, the measured analytical value includes both the added Zn and the Zn present in the filter paper. Therefore, the values are compared for Zn with this average filter blank included, to avoid an obvious discrepancy.

Table V
Toxic Metals on Filters (μg/filter)
Added/Measured Values

	Level I	Level II	Level III	Level IV (Filter Blank)
Pb	0.995	0.994	0.989	-----
Cd	0.988	0.996	0.987	-----
Zn	1.003**	1.001**	0.992**	-----
Mn	1.011	0.999	0.989	-----

**   Added value includes amount of Zn in filter

There is a consistent bias between the amount of material delivered to the filter papers and the amount measured in Level III (Added/Measured - Level I: 0.999 ± 0.010; Level II: 0.9975 ± 0.0031; Level III: 0.989 ± 0.002). The elemental determinations in a given level are not expected to be correlated since the spike concentrations were calibrated independently. The same spike mix was used in the quantitation of all four elements in all levels, and, therefore a constant difference in all levels would be expected if the measurement system were the source of this discrepancy. A standard master solution used to prepare the Level III filters which is slightly more dilute than calculated, is the most likely source of the observed bias.

Summary of Results. A summary of results of the analyses of filters from Levels I-IV are given in Table VI. The precision on the Mn measurements is about 1%. The precision on the remaining measurements as determined by isotope dilution range from 0.27% to 0.60% rsd. All four elements agree with the target amounts in the three doped levels. The calculated amount of each element at each level was in good agreement with the measured amounts within the uncertainties stated.

Figure 1. Limited Mass Range of Three Different Filter Scans

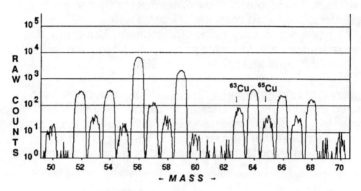

Figure 1-A. - Batch A Filter

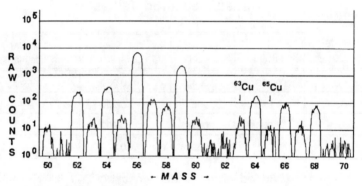

Figure 1-B. - Batch B Filter

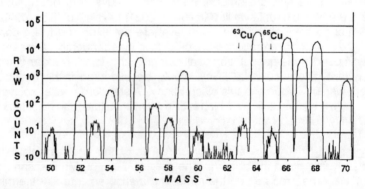

Figure 1-C. - Level III Filter

Table VI
Toxic Metals on Filters (µg/filter)
ICP-MS Measured Values

	Level I	Level II	Level III	Level IV (Filter Blank)
Pb	7.46 ± 0.03*	14.87 ± 0.09*	29.97 ± 0.12*	≤0.005
Cd	0.971 ± 0.005	2.82 ± 0.02	10.11 ± 0.05	≤0.0004
Zn	10.13 ± 0.06	49.93 ± 0.30	99.72 ± 0.27	0.26
Mn	2.08 ± 0.03	9.84 ± 0.07	19.94 ± 0.08	≤0.005

*   1s

4 CONCLUSIONS

ICP-MS was used successfully in the determination of Pb, Cd, Zn and Mn in the Toxic Metals on Filters SRM. The analytical procedures described, including quantitation by both isotope dilution and internal standard techniques, met the accuracy requirements needed in the certification of this standard reference material.

REFERENCES

1. R.S. Houk, V.A. Fassel, G.D. Flesch, and H.J. Svec, Inductively Coupled Argon Plasma as an Ion Source for Mass Spectrometric Determination of Trace Elements, Anal. Chem., **52**, 2283-2289 (1980).

2. P.J. Paulsen, E.S. Beary, D.S. Bushee and J.R. Moody, Inductively Coupled Plasma Mass Spectrometric Analysis of Ultrapure Acids, Anal. Chem., **60**, No. 10, 971-975 (1988).

3. J.R. Moody, NBS Clean Laboratories for Trace Element Analysis, Anal. Chem., **54**, No. 13, 1358A (1982).

4. J.D. Fassett and P.J. Paulsen, Isotope Dilution Mass Spectrometry for Accurate Elemental Analysis, Anal. Chem., **61**, No. 10, 643A-649A (1989).

Environmental Applications of ICP-MS Using a PQe Spectrometer

C. Vandecasteele, K. Van den Broeck, V. Dutré, and
H. Cooreman

DEPARTMENT OF CHEMICAL ENGINEERING, KATHOLIEKE
UNIVERSITEIT LEUVEN, DE CROYLAAN 46, 3001 LEUVEN, BELGIUM

1 INTRODUCTION

Several environmental applications of ICP-MS have been reported so far. Samples analysed include :
- all types of water : mineral waters, river waters, sea waters,...
- solid environmental samples : airborne particulates, fly ash, sediments, plant materials, ...

so that it is certainly impossible to review all these applications.

As an example the analysis of sea-water which is indeed a difficult matrix because of the high salt concentration and because of the low concentrations of most trace elements, will be discussed. Vanhaecke et al.(1) determined Mo in sea water using a PQ1. Sea water was tenfold diluted, In was added as internal standard and a blank containing the same Br concentration as the standard was subtracted. Table 1 shows the results. A standard deviation of ca 5% was obtained at the 100 nmol/kg (ca 10 μg/l) level (before dilution). This determination was carried out in the frame of a BCR certification campaign, so that comparison with the results from other laboratories and methods was possible. The agreement with Stripping Voltametry (SV) and Zeeman Electrothermal Atomic Absorption Spectrometry (ZET-AAS) is excellent in view of the low concentration. The concentration was therefore certified at (102.7 $\pm$ 7.7) nmol/kg. For the determination of other elements (such as Ti, V, Mn, Fe, Co, Ni, Cu, Cd, and Pb) in seawater or in high salt matrices it may be necessary to preconcentrate these elements and to remove the salt. A particularly elegant system was described by Heithmar et al. (2) , using macroporous iminodiacetate resin. Alkali and alkaline - earth metals along with e.g. chloride that can cause both matrix and spectral interferences are washed off the column and the analyte metals are then eluted with nitric acid. To date with newer versions of both low resolution (LR) and high resolution (HR) ICP-MS still lower concentrations can be determined. However, there are numerous important environmental problems where the low detection limits required for sea water are not really needed and where so far use was made of AAS or ICP-AES. The remainder of this paper will concentrate on these applications.

Table 1 : Determination of Mo in seawater with a PQ1 (1)

Method	Mean (nmol/kg)	s(nmol/kg)
ICP-MS	109.4	5.7
SV	94	11
ZET-AAS	104.4	7.3
Certified value	**102.7**	**7.7**

2 THE PQe

Recently (beginning 1991) VG-Elemental launched the PQe, an ICP-MS especially intended for environmental analyses and to meet the needs of environmental laboratories. The PQe allows rapid and full automatic sample throughput.
Figure 1 shows a scheme of the instrument.

The instrument differs from the PQ in several aspects:

- The ion optics are simplified : the PQe has only three lenses driven by a computer and no photon stop. Tuning of the lenses can be carried out via a software optimisation algorithm.

- Instead of a Channeltron detector a Faraday cup is used, connected to an autoranging amplifier, with 8 orders of magnitude dynamic range. This has of course the advantage of being non-consumable, contrary to the Channeltron which must be replaced after a period of time (e.g. 6 months depending on the applications).

- The Faraday cup does not require such a high vacuum as the Channeltron detector and does not require any preconditioning. The PQe therefore has only a 2 stage vacuum, the first stage being obtained by a mechanical vacuum pump, the second by a turbomolecular pump. In addition the vacuum can be switched off when the instrument is not in use. After a cold start the instrument is operational in about 30 minutes.

- Finally the system can only be used in the peak hopping mode, no rapid scanning is possible since there is no multichannel analyser.

Such a system is less expensive, easier to operate and less delicate than PQ systems.

A PQe has been installed in our laboratory at the Department of Chemical Engineering of K.U.Leuven in the beginning of 1991.

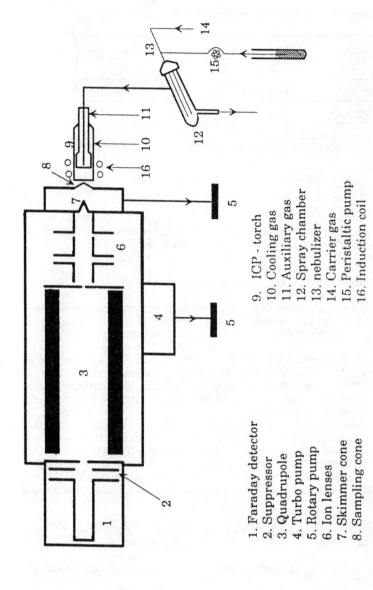

1. Faraday detector
2. Suppressor
3. Quadrupole
4. Turbo pump
5. Rotary pump
6. Ion lenses
7. Skimmer cone
8. Sampling cone

9. ICP - torch
10. Cooling gas
11. Auxiliary gas
12. Spray chamber
13. nebulizer
14. Carrier gas
15. Peristaltic pump
16. Induction coil

Figure 1: Scheme of a PQe ICP-MS (Fisons).

3 PERFORMANCE CHARACTERISTICS

Like in the PQ the signal is a function of the nebuliser gas flow and also of the ICP-power. For each power there is for a given element an optimum nebuliser gas flow and the signal increases with power (Figure 2).

The optimum gas flow seems as for the PQ to decrease somewhat with element mass, but compromise conditions can easily be found.

Table 2 summarizes the 3s-detection limits for several elements under the following conditions : 3 points per peak were measured each for 2 seconds dwell-time. A de Galan high solids nebuliser was used. Of course these limits (in general between 0.1 and 1 μg/l) are 10 - 100 times worse than those obtained by a PQ, mainly because of the less sensitive detector. They are however in general better than for ICP-AES.

The linear dynamic range is typically larger than 7 orders of magnitude, from 0.1μg/l up to 1000mg/l.

Short-term stability is in general better than 2% and long-term stability (Figure 3) better than 5% . The short-term stability test was performed by taking 10 separate 1 minute analyses over the 10 minute period. The long-term test was performed by taking 60 separate 6 minute analyses over the 360 minute period.

A study of the matrix effect due to high salt concentrations was also made. As with other commercial systems, the signal intensity in general decreases with increasing salt concentration, to an extent similar as with the PQ. We tested this for NaCl, Na_2SO_4 and Na_3PO_4. It is apparent that (in a NaCl matrix) heavy elements are suppressed more than lighter elements (Figure 4). Such suppression effect may be corrected by using an internal standard. As appears from Figure 5 for Mn 55 the correction is best when the internal standard is close in mass to the analyte element (Co and In). Selecting e.g. three internal standards : Al, In and Bi gives good corrections for all masses.

Interferences from polyatomic ions were also studied. A small list is given in Table 3. It must be stressed that for levels of e.g. Cl generally encountered in fresh water the interference is below or close to the detection limit of the procedure.

<u>Table 2</u> :   Detection limits (D.L.) of elements of environmental interest (E.P.A.) with PQe

Element	D.L. (μg/l)	Element	D.L. (μg/l)
Li 7	0.5	Nb 93	0.4
B 11	1.0	Pd 106	0.5
Na 23	10	Ag 107	0.4
K 39	14	Cd 111	0.5
Cr 52	0.4	Sn 118	0.9
Mn 55	0.3	Sb 121	1.5
Fe 56	10	Te 130	1.0
Co 59	0.7	Ba 138	0.9
Ni 60	0.7	Ta 181	0.5
Cu 65	0.5	W 184	0.8
Zn 66	1.5	Pt 198	0.8
As 75	0.7	Au 197	0.6
Se 78	4	Hg 202	-
Rb 85	1.0	Tl 205	0.1
Sr 88	0.3	Pb 208	0.1
Mo 98	1.0	Bi 209	0.3

<u>Table 3</u> : Interferences from polyatomic ions ClO and from the gas Argon

Interfering element	Interfering ion	Conc. (mg/l)	Isotope measured	Apparent conc. (μg/l)
Cl	$^{35}Cl^{16}O$	152	^{51}V	1.1
Cl	$^{37}Cl^{16}O$	152	^{53}Cr	3.1
Ar	$^{40}Ar^{16}O$	gas	^{56}Fe	75

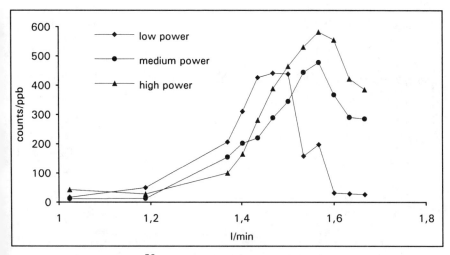

Figure 2: Signal for ^{52}Cr as a function of nebuliser gas flow for several ICP power settings.

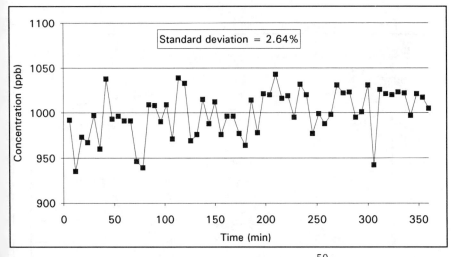

Figure 3: Long Term Stability test for a 1000ppb ^{59}Co solution.

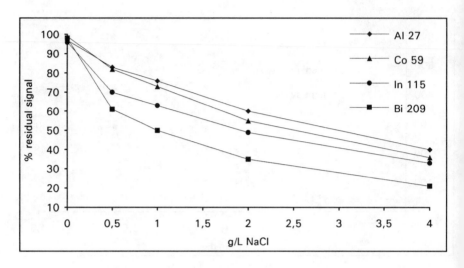

Figure 4:   Suppression of Al, Co, In and Bi in NaCl solution.

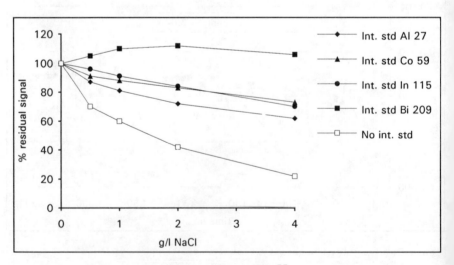

Figure 5:   Effect of internal standard for [55]Mn in NaCl.

4 APPLICATIONS

In the applications described below, the PQe was used in its standard configuration, with a de Galan nebuliser and a borosilicate spray chamber. All samples were made up in 2% nitric acid ("suprapur", Merck) and spiked with 1 mg/l Indium as internal standard. Standard solutions were made by diluting the 1 g/l standard stock solution of the individual elements (Merck).

Water Samples

In order to check if with the new instrument accurate results could be obtained we analysed several certified reference materials and participated in several round-robins, starting with water samples.

A first reference material that is frequently analysed is of course SRM 1643b from NIST, Trace Elements in Water. Table 4 gives the results : the agreement between certified and obtained values was satisfactory for all elements considered.

Table 4 : Determination of trace elements in water SRM 1643b from NIST

Element	Concentration (μg/l)	
	Found	Certified
Cr 52	18.5 ± 0.5	18.6
Mn 55	31.0 ± 0.2	28
Fe 56	95.7 ± 2.2	99
Ni 60	47.5 ± 1.1	49
Co 59	26.4 ± 0.4	26
Cu 65	20.4 ± 1.0	21.9
Zn 66	63.5 ± 2.8	66
As 75	51.5 ± 0.9	(49)
Ag 107	8.8 ± 0.3	9.8
Cd 111	19.3 ± 0.4	20
Ba 138	47.0 ± 0.3	44
Tl 205	7.5 ± 0.1	8.0
Pb 208	22.4 ± 1.0	23.7
Bi 209	9.2 ± 0.1	(11)

values between brackets are not certified.

Results were also obtained for Fresh Water CRM 398 and Fresh Water CRM 399 from BCR (Tables 5a and 5b). Again good agreement was obtained with the certified values. Analysis of this material allowed us to demonstrate that for such elements as P, and Cl at the ppm level, which are determined very often in routine water analyses, ICP-MS can yield good results.

Since our laboratory is officially recognized by our government for the analysis of waste and waste waters we were to participate on a regular basis in an intercomparison "Aquacheck" to check the accuracy and comparability between laboratories for the recognized labs. Some 40 labs in Belgium participated in the exercise. For the metals we relied systematically on ICP-MS for this exercise. Table 6 gives our ICP-MS results, and compares these with the mean values and standard deviations for the participating laboratories and with the reference value. The agreement with the reference values is in all cases excellent.

When waste water samples are analysed the large linear dynamic range of ICP-MS is a definite advantage. Concentration ranges found in waste water samples from surface treatment of metals and plastics are in general very wide (e.g. for Cr : 1.8 - 10700 mg/l and for Zn : 1.7 - 54700 mg/l). With e.g. AAS this would require preparing large numbers of standards to establish calibration curves in the appropriate concentration range.

Solid Environmental Samples

Of course we are not only interested in analysing water but also a lot of other solid environmental samples. To check the achievable accuracy several certified reference materials were analysed. Tables 7 and 8 give some results for olive leaves CRM 278 and mussel tissue CRM 62 from BCR.

We participated also in the preparation of reference materials of environmental interest. The material was prepared at BCNM, Geel and we provided some indicative values for the concentrations to be expected. The material would be used in a subsequent BGA (Bundesgesundheitsamt) round-robin in Germany. This was done for Tomato powder RM 404, Paprika powder RM 403, Lettuce powder RM 402 and Pig liver RM 005. For lettuce powder the results from the round-robin were already available. In Figures 6a and 6b our results for Copper and Lead are compared to those obtained by other analytical techniques, showing that the agreement is satisfactory. Table 9 gives the results for paprika powder and compares the results with those obtained by AAS on solution (Graphite Furnace or Flame AAS) and by AAS directly on the solid sample. Each time the agreement is satisfactory, except for cadmium where ICP-MS with the PQe lacks sensitivity.

We also apply the PQe for the analysis of membrane filters loaded with airborne dust collected in workplaces. These filters are analysed after dissolution in HNO_3.

Table 5a : Determination of trace elements in water CRM 398 from BCR

Element	This work (PQe)	certified value
Al 27 (ng/g)	40.9 ± 1.1	36.3 ± 4.3
Ca 44 (μg/g)	27.7 ± 0.4	30.0 ± 0.3
Mn 55 (ng/g)	31.6 ± 0.6	29.9 ± 0.3
Cl 35 (μg/g)	8.99 ± 0.4	10.3 ± 0.4

Table 5b : Determination of trace elements in water CRM 399 from BCR

Element	This work (PQe)	Certified value
Al 27 (ng/g)	206 ± 6	207 ± 9
Ca 44 (ng/g)	76.2 ± 2.3	79.2 ± 0.9
Mn 55 (ng/g)	192 ± 5	199 ± 3
Cl 35 (μg/g)	47.2 ± 4.2	50.5 ± 0.9
P 31 (μg/g)	1.07 ± 0.11	1.01 ± 0.03

Table 6 : Results of first round-robin of Aquacheck

Element	This work	Mean	Reference value
As (μg/l)	25.3 ± 0.5	23.6 ± 5.2	26.0
Al (mg/l)	1.83 ± 0.03	2.04 ± 0.44	1.95
Cr (mg/l)	0.101 ± 0.001	0.117 ± 0.055	0.110
Fe (mg/l)	1.76 ± 0.02	1.81 ± 0.16	1.85
Mn (mg/l)	0.474 ± 0.003	0.474 ± 0.046	0.500
Cd (μg/l)	19.5 ± 0.6	18.8 ± 4.0	19.6
Cu (mg/l)	0.232 ± 0.004	0.234 ± 0.029	0.240
Pb (mg/l)	0.096 ± 0.001	0.104 ± 0.022	0.102
Ni (mg/l)	0.075 ± 0.001	0.076 ± 0.014	0.080
Zn (mg/l)	0.613 ± 0.007	0.607 ± 0.056	0.602
Se (mg/l)	0.050 ± 0.001	0.045 ± 0.013	0.045

Table 7 : Determination of trace elements in Olea europaea CRM 62 from BCR

Element	Concentration (μg/g)	
	This work (PQe)	Certified value
Mn 55	**53.7 ± 0.4**	57.0 ± 2.4
Zn 66	**17.5 ± 0.4**	16.0 ± 0.7
Cd 111	**< 0.1**	0.10 ± 0.02
Pb 208	**25.6 ± 0.4**	25.0 ± 1.5

Table 8 : Determination of trace elements in mussel tissue CRM 287 from BCR.

Element	Concentration (μg/g)	
	This work (PQe)	Certified value
Mn 55	**7.6 ± 0.1**	7.3 ± 0.2
Fe 56	**134 ± 2**	133 ± 4
Cu 65	**9.52 ± 0.15**	9.60 ± 0.16
Pb 208	**1.99 ± 0.01**	1.91 ± 0.04

Table 9 :   Comparison of three methods for the determination of trace elements in Paprika powder CBNM-RM 403.

Element	Concentration (μg/l)		
	SS-AAS	AAS	**This work (PQe)**
Cd	0.55 ± 0.10	0.41 ± 0.02*	**< 0.5**
Cr	-	0.6 ± 0.1* 0.3 ± 0.1	**0.4 ± 0.1**
Cu	9.4 ± 0.9	10.8 ± 0.1	**10.3 ± 0.4**
Pb	0.08 ± 0.01	0.08 ± 0.02*	**0.09 ± 0.05**
Tl	-	-	**< 0.01**
Zn	19.9 ± 0.8	18.1 ± 1.3	**20.1 ± 1.7**

* Graphite Furnace AAS, other figures are Flame AAS.

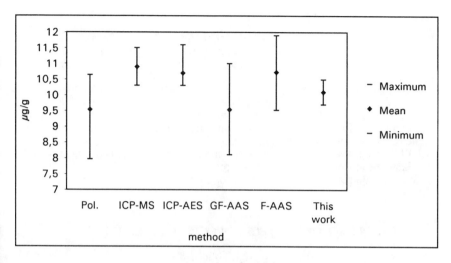

Figure 6a: Results of BGA round-robin for Copper in Lettuce powder RM 402.

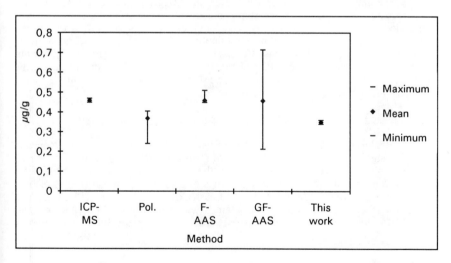

Figure 6b: Results of BGA round-robin for Lead in Lettuce powder RM 402.

5 LIMITATIONS

In practice for many real samples detection limits for As, Se and Hg are not sufficient.   It goes without saying that these elements are extremely important for environmental purposes. Therefore, recently we acquired a hydride generator to be linked to the PQe. (3)

Also in general from time to time somewhat better detection limits are needed. In practice this is the case e.g. for water samples with a low matrix concentration. In such cases using an ultrasonic nebuliser may be of much use. Table 10 gives the detection limits that can be achieved with such a device (CETAC, with desolvation), they are on average 10 times better than with the de Galan nebuliser.   Table 11 gives the results obtained for SPA REINE, a commercial Belgian drinking water. For most elements precise results can be obtained.

6 APPLICATIONS IN TECHNOLOGICAL RESEARCH

Study of Leaching of Metals (As) from Solidified Waste.

We have applied the PQe spectrometer in several technological research projects. The first example is the study of the leaching of heavy metals, including arsenic, from solidified waste. Waste containing arsenic was submitted to leaching tests (1 gram of waste in 10 ml of water was shaken for 24 hours). The waste material was then solidified as follows : per gram of waste 1.5 gram slags, 0.5 gram of iron containing acid and water were added in order to obtain a slurry (1st step). One day later 1.1 gram of cement and 1 gram of lime were added plus some water (2nd step). The mixture was then allowed to solidify for 1 week.

The amount of arsenic leached after this two step procedure was only 5.2 mg/l instead of 4.2 g/l. The two step process was evaluated and the influence of the added components was investigated : addition of slags, acid, or iron (in the acid) does not have an influence on the amount leached. Only lime seems to be an important factor, probably because $Ca_3 (AsO_4)_2$ is hardly soluble (Figure 7). At the same time the leachability of other elements (Sb, Pb, Zn) from the waste can be measured simultaneously. In addition, a one step process was tested, which seemed to give similar results to the two step process.

Leaching of Heavy Metals from Fly Ash from Municipal Incineration.

The leaching of heavy metals from fly ash from municipal waste incineration was also studied. 26 or 14 elements were measured simultaneously. Figures 8a and 8b show the extraction behaviour i. e. the amount extracted (mg per kg of fly ash as a function of the acid dose, i.e. moles of acid per kg of ash (moles of base if dose is negative). Cadmium is leached easily even with a low acid dose. The amount leached is compared to that obtained with aqua regia. The yield is about 80% for 4 mol/kg of ash. Copper is only leached to a reasonable extent for very high acid doses.

Table 10 :   Detection limits (ppb) with an ultrasonic nebuliser

Element	Ultrasonic nebuliser	de Galan nebuliser	Element	Ultrasonic nebuliser	de Galan nebuliser
Be 9	0.07	0.07	Cu 65	0.09	0.5
Na 23	1.6	10	Zn 66	0.3	1.5
Mg 24	0.1	0.3	As 75	0.9	0.7
Al 27	0.4	0.5	Se 78	1.5	4
K 39	5	14	Cd 111	0.3	0.5
Ca 44	2.8	32	Sn 118	0.1	0.9
Cr 52	0.03	0.4	Sb 121	0.2	1.5
Mn 55	0.01	0.3	Ba 138	0.02	0.9
Fe 56	0.4	10	Tl 205	0.01	0.1
Co 59	0.03	0.7	Pb 208	0.02	0.1
Ni 60	0.06	0.7			

Table 11 :   Concentration (μg/l) of minor elements in SPA REINE (undiluted) with an ultrasonic nebuliser

Element	Mean value (*)
Mn 55	38.8 $\pm$ 0.6
Fe 56	5.09 $\pm$ 0.45
Co 59	1.45 $\pm$ 0.03
Ni 60	2.14 $\pm$ 0.12
Cu 65	0.55 $\pm$ 0.06
Zn 66	58.7 $\pm$ 0.8
Cd 111	< 0.25
Tl 205	< 0.01
Pb 208	< 0.02

(*) Mean value of 10 repeats

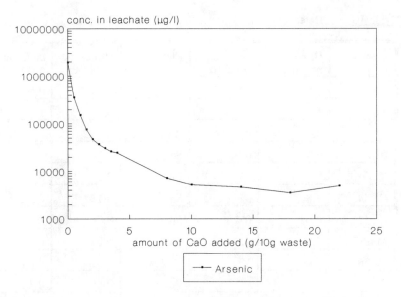

Figure 7:   Leachability of arsenic in function of CaO added.

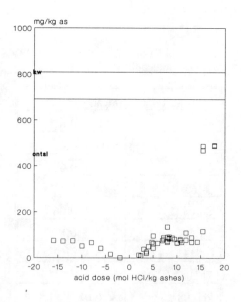

Figure 8a: Amount of Copper extracted from fly ash as a function
             of the acid dose.

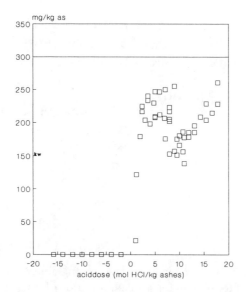

Figure 8b: Amount of Cadmium extracted from fly ash as a function of the acid dose.

7 CONCLUSION

ICP-MS   is a low cost, low resolution instrument can be used with success in many environmental applications.

REFERENCES

(1)   Vanhaecke F., Goossens J., Dams R. and Vandecasteele C., submitted for publication.

(2)   Heithmar E.M., Hinners T.A. and Henshaw J. M., Anal. Chem., 1989, 61, 335.

(3)   Hitchen P., Hutton R. and Tye C, J. Autom. Chemistry, 1992, 14, 1, 17.

Investigation Into the Feasibility of ICP-MS as an Alternative to Fire Assay Measurements for Gold and the Platinum Group Elements

J. Godfrey and E. McCurdy

VG ELEMENTAL LTD, ION PATH, ROAD THREE, WINSFORD, CHESHIRE
CW7 3BX, UK

Introduction

The platinum and gold mining industry has for some time been interested in the development of techniques for the determination of Au and the platinum group elements (PGE's) in geological samples. Due to the very low concentrations of precious metals occurring in geological samples, Fire Assay has historically been the only convenient method of collecting and concentrating these elements for analysis[1]. The process involves the use of dry reagents such as lead oxide, silica, borax and sodium carbonate, followed by heating. There are three basic steps to the process:-

* Ore is mixed with the above flux and a small amount of a reducing agent such as carbon. This mixture is placed in a crucible and loaded into a muffle furnace at 1100°C for approximately 30 minutes. The flux reacts with the ore, and the entire mixture becomes molten. Carbon reacts with some of the lead oxide reducing it to lead. Due to the differences in specific gravity between the lead and the slag remaining, two layers form. The large solubility of PGE's and Au in lead, and their almost complete insolubility in the slag, forms the basis of the separation. After cooling, the slag is removed and the resulting lead button containing the precious metals remains.

* The lead button is then subjected to a cupellation process in which the lead is oxidized and separated from the precious metals. The lead button is placed in a pre-heated porous dish called a cupel. The cupel is placed in a muffle furnace at 1000°C, with sufficient airflow to oxidize the lead. As the lead oxidizes, it soaks into the cupel, draining with it any other base-metal oxides that may be present. The PGE's and Au do not oxidize and remain as a small round bead on top of the cupel.

* The PGE's and Au are then dissolved and the solution analysed by plasma OES or atomic absorption spectrometry.

Despite the widespread use of fire assay techniques, these procedures have some drawbacks. The perceived failings are the extensive sample preparation requiring a great deal of skill. The technique is time consuming, labour intensive and also very costly.

With these drawbacks in mind, it was decided to evaluate an alternative procedure to the Fire Assay technique. Due to the exceptional sensitivity and dynamic range of Inductively Coupled Plasma Mass Spectrometry (ICP-MS) it is capable of extending applications to samples containing parts per trillion (ppt) levels of many elements including the PGE's (see Table 1). Therefore, a much simpler fusion procedure was developed, allowing direct analysis of the digested fusion bead by ICP-MS.

Ru	12
Rh	1
Pd	25
Ir	4
Pt	7
Au	6

Table 1 - Detection Limits for PGE's and Gold (ng.l⁻¹)

A schematic flow chart illustrating several possible sample preparation and analysis techniques for PGE's and Au determinations in rocks and minerals is shown in Figure 1.

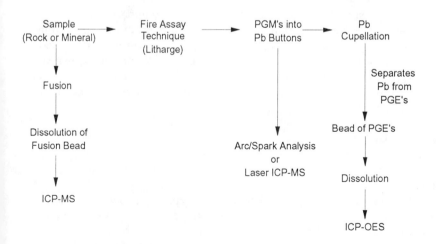

Figure 1 - Schematic Sample Preparation Flow Chart

Experimental
The samples used for these experiments were obtained from the Merensky Reef, South Africa, from which the main concentration of PGE's are mined.

Fusion Process for ICP-MS Analysis

Method 1

Approximately 1 gram of sample (mineral/ore) and 5 grams of sodium peroxide were weighed accurately and fused in a zirconium crucible. The fusion bead was then dissolved in 50mls of concentrated hydrochloric acid and then diluted to 250ml with deionised water (18 Mohm). The samples were all prepared in triplicate.

Two blanks were prepared by fusing 5 grams sodium peroxide in zirconium crucibles and dissolving the fusion bead in 50mls of hydrochloric acid. This mixture was then diluted to 250ml with deionised water, to give a matrix blank of similar composition to the unknown samples.

Method 2

0.5 gram of sample was fused in a zirconium crucible with 4 grams of sodium peroxide. The fusion was leached in 40ml of deionised water. The solution was heated with 40ml of HCl, boiled to remove excess peroxide, cooled and diluted to 100ml with deionised water.

A fusion blank was prepared using the fusion/leaching procedure as described above, but with no sample.

Sample Dilution
When preparing solutions for analysis by ICP-MS, an important consideration is the concentration of total dissolved solids (TDS) present in the solution when analysed. As a general rule, this concentration should not exceed 0.25% (2500 $\mu g.ml^{-1}$) as, above this level, sample cone blockage may occur, which can lead to severe analyte signal drift. A suitable further dilution factor for the samples analysed in this work would be 10x dilution for Method 1 and 20x dilution for Method 2.

Sample Analysis
In order to match the acid matrix of the sample solutions, the calibration blank and multi-element calibration standard solutions were prepared in 2% HCl. 20ppb of Caesium and Bismuth were added to all samples and standards as internal standards. The samples prepared by Method 1 were analysed this way. Samples prepared by the fusion Method 2, were analysed against synthetic standards prepared by spiking the fusion blank solution at various concentrations of PGE's and Au. Again, an internal standard was added to all of the samples and standards.

All samples were analysed using a PlasmaQuad PQ2 (VG Elemental, Winsford, Cheshire, UK) Details of the instruments parameters are given in Table 2.

Plasma forward power	1350W
Plasma reflected power	0W
Coolant gas flow	13.5 l.min^{-1}
Auxiliary gas flow	0.5 l.min^{-1}
Nebuliser (DeGalen V-groove) gas flow	0.90 l.min^{-1}
Sample uptake rate	0.8 ml.min^{-1}
Ion lens settings optimised on	^{115}In

Table 2 - Instrumental Operating Conditions

Full mass scans were performed on two of the samples to give full spectral information. In this way, unexpected elements can be identified and the isotopic fingerprints of elements can be checked for unambiguous sample characterisation. For example, a high concentration of zirconium was identified in these samples (from the crucibles), which can cause an interference from ZrO^+ on some of the Pd isotopes (See Figure 2). Also HfO^+, which overlaps some of the Pt isotopes, was identified from the scan spectrum. Using the information on potential interferences, obtained from the full mass scans, care was taken in selecting uninterfered isotopes of Pd and Pt before acquiring data by Peak Jumping.

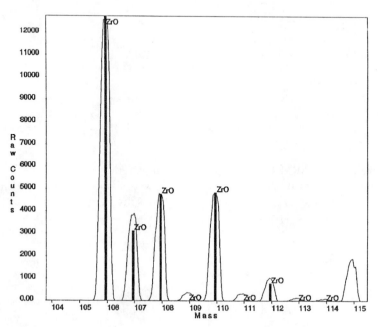

Figure 2 - PlasmaQuad Scan of Blank Fusion Matrix, Showing Identification of ZrO$^+$ Peaks

Sample Analysis using Flow Injection ICP-MS

Flow Injection ICP-MS was used as an alternative technique to continuous nebulisation, in order to avoid the dilution step which is required for successful analysis of samples containing high dissolved solids, when using continuous nebulisation. The use of Flow Injection ICP-MS enables the samples to be run directly at relatively high dissolved solids levels, hence reducing any sample preparation contamination problems and potentially improving the detection limits. Since smaller sample volumes were aspirated, analysis of solutions containing high concentrations of dissolved solids could be performed without significant matrix deposition on the ICP-MS interface.

Results and Discussions

Table 3 shows the results obtained in this study, for the PGE's and Au in the standard reference material SARM 7 and a series of Merensky Reef PGE-ore samples. Mean concentrations and standard deviations for the samples are shown. These samples were prepared by Method 1 and calibrated using synthetic standards, SARM 7 reference values are also shown for comparison.

	Ru(101)	Rh(103)	Pd(105)	Ir(193)	Pt(194)	Pt(195)	Au(197)
WFF Tail	0.231	0.208	0.520	0.173	1.272	1.321	0.646
RSD	(0.017)	(0.012)	(0.015)	(0.004)	(0.044)	(0.024)	(0.126)
Furnace Matte	97.49	73.89	250.0	23.21	449.7	448.8	37.09
RSD	(1.027)	(0.500)	(1.584)	(0.210)	(6.307)	(6.377)	(2.069)
Converter Matte	253.3	225.9	592.7	76.86	1073	1070	134.5
RSD	(0.226)	(0.855)	(1.376)	(0.501)	(4.308)	(7.529)	(6.785)
SARM 7	0.435	0.933	1.473	0.175	3.505	3.365	0.515
RSD	(0.021)	(0.008)	(0.036)	(0.004)	(0.012)	(0.018)	(0.041)
SARM 7 Ref Values	0.430	0.240	1.530	0.074	3.740	3.740	0.310

Table 3 -Quantitative Results for Merensky Reef Samples and SARM 7 Prepared Using Fusion Method 1. All Values μg/g n=3.

The results for SARM 7 show good agreement with the reference values, for all elements except Ir, Rh and Au, all of which were significantly higher than the reference values. The discrepancies in the results are thought to be due to spectral interferences of:

Rh[103] -   ArCu
Ir[193] -   HfO
Au[197] -   TaO

Whilst the ArCu peak is due to the presence of copper in the ore samples, the latter two interferences are due to high concentrations of elements associated with the zirconium crucibles used for the fusion. An alternative crucible material would reduce these interferences.

Despite the interferences present on some elements, it is clear that ICP-MS analysis can give excellent precision for the PGE's at these concentrations.

As described earlier, the sample solutions prepared by the Fusion method contain high total dissolved solids, so a dilution is required before the solutions can be analysed by continuous nebulisation. To avoid this dilution step, the samples prepared by Method 2 were run using Flow Injection ICP-MS, giving the following advantages:

* Direct analysis of high dissolved solids
* Increased sample throughput over continuous nebulisation
* Full multi-element analysis from small sample volumes

The samples SARM 7 and Final Concentrate were calibrated using matrix matched standards, to reduce the effect of the fusion matrix. Results for the Flow Injection ICP-MS analysis of these samples are shown in Table 4.

	Ru	Rh	Pd	Ir	Pt	Au
SARM 7	0.373	0.204	1.799	0.159	3.610	0.320
Reference	0.43	0.24	1.53	0.074	3.74	0.31
Final Conc	10.54	6.58	30.42	1.93	65.1	3.16
Reference	9.5	6.1	29.2	1.8	62.5	3.1

Table 4 - Flow Injection ICP-MS Results for Fusion Samples (Method 2) - Using Matrix Matched Standards. All Values μg/g.

The data show that by using Flow Injection ICP-MS and matrix matched standards, the results for all elements in SARM 7 compare well with the reference values. This indicates that the use of matrix matched standards was effective at overcoming the polyatomic ion interferences caused by the matrix elements, which were discussed earlier. However, SARM 7 and the Merensky Reef samples indicated in tables 3 and 4 contain high concentrations of platinum group elements. For lower concentrations of the PGE's and Au, even the detection power of Flow Injection ICP-MS may be insufficient, without some form of preconcentration.

One further consideration is that the PGE's and Au tend not to be distributed uniformly throughout a bulk ore sample.[2,3] Thus the sample preparation and, particularly, the particle size present after crushing and milling, may have a profound effect on the homogeneity of the final sample. If a mean particle size of between 75 and 100 μm is assumed and a PGE concentration of 5 grams per tonne (5 μg/g) is used, then it has been estimated that all of the PGE's in a 0.5 gram sample may be contained in a single particle.[2,3]

Clearly, sub samples of 0.5 gram may, therefore, be very variable. From the graph in Figure 3, it can be seen that an acceptable number of PGE-containing particles is only obtained by increasing the sample weight (e.g. to 100 grams), or reducing the particle size (preferably to below 50 μm diameter) or both.

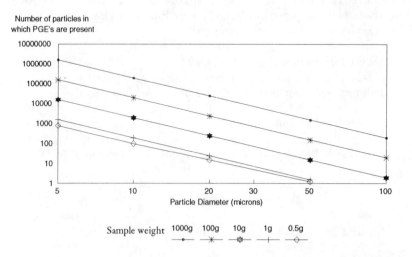

Figure 3 - Number of Particles Containing PGE's in a Sample, for Different Particle Diameters

SARM 7, which is ground to much smaller particle sizes than production ore samples, is an ideal case which is not achieved in routine preparations. Thus, SARM 7 may not be ideal as a test sample for evaluating sample preparation and analysis techniques for PGE-containing ore samples, unless the ore samples can be ground to much smaller particle size.

Conclusions and Future Work

Direct analysis of fused minerals by solution ICP-MS has been shown to give good sensitivity and precision. Good results for SARM 7 and samples containing relatively high concentrations of PGE's were obtained, and for these samples this would be a cost effective method. Flow Injection ICP-MS is the preferred technique for the analysis of samples which contain high dissolved solids, as sample preparation is reduced, detection limits are improved and sample throughput is increased.

For this to be a viable method for measuring low concentrations of PGE's in production samples, further work is required into the development of a sample preparation procedure to ensure sample homogeneity. For example the use of larger sample weights (50-200g) or smaller particle sizes is recommended.

Use of a crucible material other than zirconium may be effective at reducing certain polyatomic interferences on some of the PGE's.

References

1.   V.I. Bennet, 'Precious Metal Analysis'. ASTM Standardisation News, 1988, Dec., 62

2.   C.A. Cousins, Johannesburg Consolidated Investment Company Ltd, Platinum Group Metals, p. 94

3.   E. R. Schmidt, The Structure and Composition of the Merensky Reef and Associated Rocks in the Rustenburg Platinum Mine. <u>Trans. Geol. Soc.</u>, 1952, S. Afr., 55

Application of Flow Injection Analysis to ICP-MS

S. T. G. Anderson, M. J. C. Taylor, and S. J. S. Williams
ANALYTICAL SCIENCE DIVISION, MINTEK, PRIVATE BAG X3015,
RANDBURG 2125, SOUTH AFRICA

1 INTRODUCTION

The technique of flow injection analysis (FIA) is based on the reproducible injection of a well-defined volume of sample into a carrier stream, which carries the sample to a detector.[1] The result is a transient signal, the size and shape of which are dependent on the sample volume and the carrier flowrate. In conventional inductively coupled plasma-mass spectrometry (ICP-MS) analysis, the sample is aspirated directly into the system and the signal is measured at steady state. By combining flow injection with ICP-MS, several features of flow injection can be used to improve ICP-MS analysis. These features include increased sample throughput, reduced sample volumes, and automation of sample preparation.

Figure 1 is a diagram of a flow-injection system in its simplest configuration. The sample and carrier streams are fed into the flow-injection valve by a peristaltic pump. The valve can be switched between two positions, 'load' and 'inject'. In the 'load' position, the sample passes through the sample loop and runs to waste, while the carrier solution passes through to the detector. In the 'inject' position, the two streams are switched so that the carrier stream pushes the contents of the sample loop to the detector.

Figure 2 shows a comparison of flow-injection and steady-state signals detected by ICP-MS. For multi-element measurement of a transient signal, the detection should ideally be simultaneous. Although ICP-MS detection is strictly speaking scanning rather than simultaneous, the quadrupole scan rates are so fast that essentially simultaneous detection is achieved. Flow injection is therefore well suited to ICP-MS analysis.

2 APPARATUS

The flow-injection system was designed for use with a VG Plasmaquad II+ ICP-MS instrument equipped with a Gilson Model 222 sample changer. The

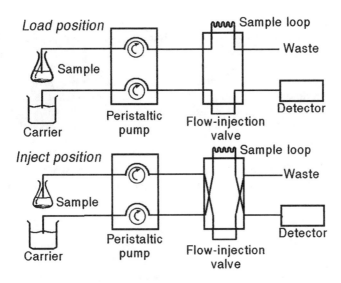

Figure 1　A SIMPLE FLOW-INJECTION SYSTEM

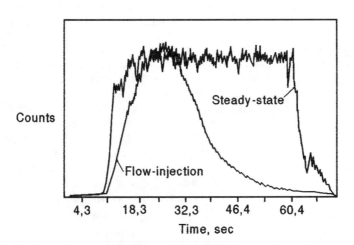

Figure 2　COMPARISON OF FLOW-INJECTION AND STEADY-STATE SIGNALS FOR 100 µg/l In

system consists of a Gilson four-channel peristaltic pump, a Dionex eight-port pneumatic flow- injection valve, and an electronic control circuit. The valve is driven by compressed gas. The switching can be done manually or electronically in combination with the autosampler. Electronic switching is achieved by means of a microswitch fixed to the autosampler arm, which is triggered each time the autosampler moves from one sample to the next. The microswitch in turn triggers a solenoid valve that switches the gas flow to the pneumatic flow-injection valve and sets the valve to the 'load' position. An adjustable timer switch allows sufficient time for the sample to fill the sample loop and then switches the valve to the 'inject' position. The uptake delay time in the Plasmaquad software program is used to synchronize the detection of the flow- injection signal with the acquisition of data by the instrument.

3 EXPERIMENTAL

Reduction of Analysis Time

When an instrument is required to perform a great deal of routine analysis, anything that enables a greater sample throughput is very advantageous. This is particularly true of ICP-MS, where the high cost of the instrumentation prevents the purchase of several instruments. Reducing the time required for routine analysis also frees the instrument for research applications. Shorter analysis times are mainly achieved by reducing the sample loading of the instrument and thereby reducing the wash time between samples. In steady state analysis, several millilitres of sample are aspirated into the instrument per analysis and considerably more time is required for the sample to reach steady state in the plasma and to wash out the system than is required for the acquisition of analytical data. By contrast, in FIA, sample volumes are typically of the order of a few hundred microlitres, and the wash time is much shorter. Obviously, as the sample volume is reduced, the wash time gets shorter, but there is also a loss of sensitivity, and it is necessary to select a compromise sample loop size that gives a short wash time while maintaining adequate sensitivity. A number of different loop sizes were tested, and it was found that with a 350 μl sample loop it was possible to achieve 70% of normal sensitivity while shortening the analysis time per sample by 40%.

One way of achieving the best possible sensitivity and shortest wash time with a flow-injection system is to minimize dispersion of the sample plug in the tubing. Figure 3 shows the behaviour of a sample plug from the moment of injection. Initially the plug is cylindrical, but due to the fact that the flowrate is highest at the centre of the tube and tends towards zero at the walls, the plug is dispersed as it flows through the tube. Therefore, the longer the tube between the injection valve and the nebulizer, the broader and flatter the time-resolved signal. This results in lower sensitivity and longer wash times. The solution to this problem is to position the injection valve as close as possible to the nebulizer, and it was found that by placing the valve

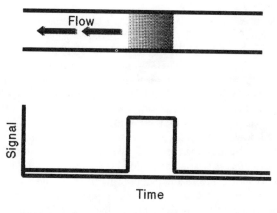

Figure 3   SAMPLE DISPERSION IN A TUBE

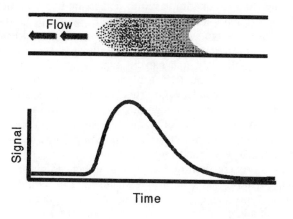

Figure 3a   SAMPLE DISPERSION IN A TUBE

inside the torch box with only a few centimetres of tubing connecting the valve and the nebulizer, it is possible to obtain better sensitivity and shorter wash times. However, some dispersion still occurs in the spray chamber. A direct-injection nebuliser system, a modified version of the system described by Wiederin, Smith, and Houk,[2] is currently being assembled at Mintek. This should reduce signal dispersion even further.

High Dissolved-Solids Analysis

The second major advantage of flow injection as a sample introduction method for ICP-MS is that it allows higher levels of dissolved solids in sample solutions. In steady-state analysis the generally accepted upper limit for dissolved solids in a sample solution is 0.2%. Higher levels result in blockages of the cone interface. Acid concentrations are usually restricted to 2%, since higher levels cause rapid deterioration of the cones. The result of this is that the analysis of solid materials inevitably involves large dilution factors and therefore raised detection limits. This is particularly serious for samples that require dissolution by alkaline fusion. These samples are typically diluted 10 000 times before analysis, so that a detection limit of 1 μg.l^{-1} in solution equates to 10 μg.g^{-1} in the original material. In flow injection analysis, the volume of sample aspirated per analysis is typically an order of magnitude lower than the volume per steady state analysis, and therefore the sample matrix and acid concentration can be an order of magnitude higher without increasing the rate of cone blockage or wear. In this way lower detection limits can be achieved.

Trace Impurities in Tantalum Oxide.   At Mintek, the use of flow injection for the analysis of samples with a high matrix content has been applied to the determination of trace impurities in high-purity tantalum oxide. This analysis is done to monitor the purification process and to certify the final product as better than 99.99% pure. Quantitative dissolution of tantalum oxide is extremely difficult, but it can be achieved by fusion with a large excess of potassium pyrosulphate in a platinum crucible and dissolving the melt in dilute hydrofluoric acid. Using steady-state analysis, a large dilution would be necessary because of the high flux-to-sample ratio and the presence of hydrofluoric acid. However, dilution must be kept to a minimum because of the required lower limits of determination. Using flow injection with 350 μl sample volumes, it was found that it is possible to analyse solutions that are five times more concentrated than the upper limit for steady-state analysis. Figure 4 shows the system that was used. Two sample loops were used, each 350 μl in volume. The advantage of this is that injection of the one loop and loading of the other loop with the next sample occur simultaneously, which saves time.

The samples were measured against matrix-matched standards with a calibration range of 0.5 to 50 μg.l^{-1}, corresponding to a range of 1.25 ppm to 125 ppm in the samples. The instrumental conditions that were used are shown in Table 1. The peak-jumping mode of data acquisition was used, because the analyte isotopes were spread over a wide mass range. The acquisition time was selected to give maximum sensitivity and precision for the selected loop size.

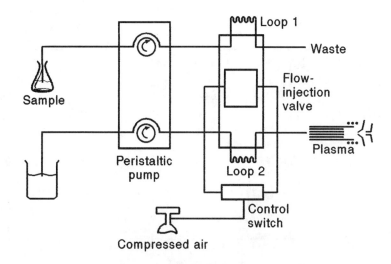

Figure 4 FLOW–INJECTION MANIFOLD FOR HIGH-SOLIDS ANALYSIS

Table 1   Operating conditions for the determination of trace impurities in tantalum oxide by FIA-ICP-MS

Loop size	350 μl
Carrier-solution flowrate	0.7 ml/min
RF power	1.35 kW
Analysis mode	Peak jump
Detector	Pulse count
Mass-range sweeps	25
Injection delay	15 s
Acquisition time	20 s

The results of repeated analysis of a tantalum oxide sample are shown in Table 2. The concentrations are in ppm in the original sample. The limits of detection are based on three times the relative standard deviation of a reagent blank, taking the dilution factor into account. The relative standard deviations are all in the region of 5%. This level of precision is considerably better than that achieved for other techniques that have been used at Mintek to analyse tantalum oxide samples, including laser ablation ICP-MS and d.c. arc-emission spectrography. After a lengthy analysis procedure involving

Table 2   Results obtained for impurities in tantalum oxide by FIA-ICP-
               MS

Element	Isotope	Concentration (μg.g^{-1})	RSD (%)	Limit of detection (μg.g^{-1})
Mg	24	1900	5.7	10
Mn	55	72	5.0	6
Nb	93	1600	3.3	0.7
Ag	107	31	2.2	2.8
Sn	120	21	2.6	6.8
W	184	17	4.1	1.4

about 50 injections, the cones showed no sign of abnormal build-up or wear. It should therefore be possible to analyse even more concentrated solutions and thereby achieve better detection limits.

On-Line Sample Manipulation

As has already been described, flow injection can be used simply as a means of sample introduction for ICP-MS. As such, flow injection is able to increase sample throughput and increase the maximum allowable matrix levels in samples. However, possibly the greatest capability of the technique is in the area of automation of sample preparation by on-line methods.

Automation of analysis has two main advantages: it reduces manpower costs and improves the reliability of results by minimizing the likelihood of analyst error. Computerized data processing and the use of autosamplers has to a large extent achieved automation of the analysis of samples. However, the preparation of samples prior to analysis is still mainly done by hand, a task which is tedious and repetitive. There are several aspects to sample preparation for ICP-MS analysis. Solid samples must be dissolved, solutions must be diluted to suitable levels of matrix or analyte elements, internal standards are usually added to correct for instrument drift and matrix effects, and in some cases it is necessary to separate analyte elements from interfering elements or to preconcentrate analyte elements prior to analysis. Of these various aspects, sample dissolution is the most difficult to automate by incorporation into an on-line system. Some work has been published on the use of on-line microwave and pressure dissolution, but it is not applicable to the sample types analysed at Mintek. However, there is potential for the automation of all other aspects of sample preparation by using flow injection.

On-Line Dilution and Internal Standard Addition

The simplest tasks to automate are dilution and internal standard addition. Automation can be achieved using the system shown in Figure 5. The system uses a mixing tee which merges the sample with a stream containing diluent and internal standard. Mixing tees are made by Upchurch, and are designed to create turbulent flow in a very small dead volume at the point of merging, so that efficient mixing of the two streams is achieved. A limitation of this on-line dilution system is that the dilution factor cannot be altered without changing the pump tubing.

An alternative system is shown in Figure 6. In this manifold the sample and diluent streams are merged. The merged stream is then split and a part of the stream is pumped to the injection valve while the excess runs to waste. The diluted stream is merged with the internal standard stream before loading the sample loop. This system is able to achieve larger dilution factors (up to approximately 250 fold) without exceeding the flowrate specifications of the mixing tees. Another advantage of this system is that it is possible to select one of two different dilution factors without changing the pump tubing. Using this system, many samples which are received in solution form can be analysed as received, with no sample preparation at all.

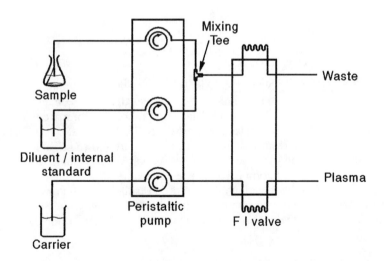

Figure 5   MANIFOLD FOR ON-LINE DILUTION OR INTERNAL
STANDARD ADDITION (1)

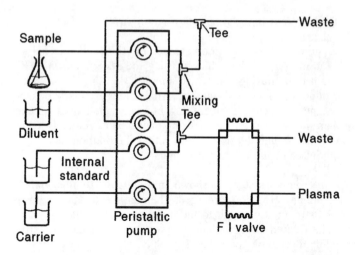

<u>Figure 6</u>   **MANIFOLD FOR ON-LINE DILUTION AND INTERNAL STANDARD ADDITION (2)**

<u>Platinum-group Metal Analysis</u>. The on-line dilution system was tested on a set of platinum-group metal (PGM) samples that had been analysed in the conventional way. The samples included a solution prepared from a number of nickel sulphide collections on SARM 7, which is used as a quality-control sample. The results obtained for SARM 7 are shown in Table 3. The agreement with the accepted values is good for all elements except palladium, and the precision is good. The high value for palladium is due to a high blank resulting from high concentration palladium solutions that had been analysed previously using conventional nebulisation. Good agreement with the results of the conventional analysis was also obtained for the other samples.

<u>Rare-earth element analysis</u>. The system has also been used for rare-earth analysis. The results obtained for the analysis of an in-house quality-control sample containing the rare earths are shown in Table 4. The results are in good agreement with the accepted values, and good precision was achieved.

Table 3   Results of analysis of SARM 7 by FIA-ICP-MS using on-line
dilution and internal standard addition

Element	Isotope	Concentration $(\mu g.g^{-1})$	RSD (%)	Accepted value $(\mu g.g^{-1})$
Pt	195	3.77	1.8	3.74
Pd	105	1.81	8.8	1.53
Ru	99	0.44	5.5	0.43
Rh	103	0.24	4.2	0.24
Ir	193	0.089	11.7	0.074
Au	197	0.33	4.6	0.31

Table 4   Results of analysis of rare-earth elements by FIA-ICP-MS using
on-line dilution and internal standard addition

Element	Isotope	Concentration $(\mu g.g^{-1})$	RSD (%)	Accepted value $(\mu g.g^{-1})$
La	139	0.56	1.2	0.54
Ce	140	1.34	0.8	1.32
Sm	152	0.14	3.0	0.14
Pr	141	0.19	1.6	0.19
Y	89	0.11	1.8	0.11
Gd	157	0.10	3.4	0.10

On-Line Ion-exchange Separation

Another aspect of automation that is being investigated is separation
and preconcentration using on-line ion-exchange columns. These operations
are more commonly performed prior to ICP emission analysis, where spectral
interferences are more severe and detection limits are higher, but there is
also potential for use with ICP-MS, particularly in the separation of trace
analytes from a more concentrated matrix. The ion exchange columns are
manufactured at Mintek; they are made of Perspex and are compatible with
standard flow-injection fittings.

In order to achieve efficient separations it is necessary to have sufficient contact between the sample and the resin. However, there are limitations on the resin-particle size and column length due to the back pressure of the column, which can become too high for the peristaltic pump.

PGM Separations.   An application of ion chromatography, which is currently being investigated, is the separation of the PGMs from base metals. It is often necessary to analyse trace levels of the PGMs in the presence of high levels of base metals. This generally requires dilution of the concentrated matrix, which results in degradation of detection limits for the PGMs. There are also potential interferences due to the overlap of polyatomic ions, for example $^{63}Cu^{40}Ar$ on the single ^{103}Rh isotope. The separation of the PGMs from base metals is therefore advantageous. There is an off-line separation method in use at Mintek, based on the loading of base metals onto a cationic resin (Dowex 50W-X8 hydrogen form). The anionic chloro-complexes of the PGMs pass directly through the column. This method has been successfully incorporated into an on-line separation system. A potentially more useful approach would be to load the PGMs onto an anion-exchange resin, thereby preconcentrating the trace-analyte elements while simultaneously removing the base-metal matrix. This has been attempted with a resin called Monivex, which was developed at Mintek. However, it was found that, while the PGMs loaded efficiently onto the resin, extreme conditions (for example brominated hydrochloric acid) were required to strip them off again for analysis. This method was therefore not suited to an on-line system.   Other resins are to be investigated. There is also the possibility of incorporating other separation methods such as solvent extraction in an on-line system.

4   CONCLUSION

To sum up, FIA is an attractive alternative to conventional sample introduction for ICP-MS, and can be used on a routine basis. Flow injection can be used to increase the sample throughput rate and to increase the matrix tolerance of the ICP-MS system. Flow injection also facilitates on-line sample preparation and manipulation, and there are still a number of possibilities to be investigated in this area.

REFERENCES

1.   J. Ruzicka and E.H. Hansen, 'Flow Injection Analysis', John Wiley and Sons, New York, 2nd Edition, 1988, Chapter 2, p. 15.

2.   D.R.Wiederin, F.G.Smith and R.S.Houk, Anal. Chem. 1991, 63, 219.

Decomposition Temperature and its Influence on Trace Element Determination by ICP-MS and ICP-AES

P. Fecher, M. Leibenzeder, and C. Zizek
LANDESUNTERSUCHUNGSAMT FÜR DAS GESUNDHEITSWESEN
NORDBAYERN, HENKESTRASSE 9-11, D-8520 ERLANGEN, GERMANY

1 INTRODUCTION

Determination of trace elements in biological materials needs a destruction of organic matrix to bring the sample to a liquid form, suitable for analysis with e.g. ICP-Mass-Spectrometry (ICP-MS) or ICP-Atomic-Emission-Spectrometry (ICP-AES). In special cases liquid samples like serum, blood, milk or juice can be analysed directly or with a dilution step, using the high temperature of the ICP as an ashing agent itself. Depending on the type of sample, nebulisation effects, interferences or even carbon depositions on the torch can be observed in these cases. Therefore an ashing step prior to determination is essential to achieve accurate results.

Among the different types of decomposition for destruction of organic matrix[1], digestion with nitric acid in closed vessels is the most suitable procedure for trace element determinations. Advantages of pressure digestion are: vessels made of quartz or fluorpolymer, no volatilisation of elements, no contamination from external sources, fast digestion and good suitability for routine use.

Depending on the type of instruments and operating parameters, different qualities of decomposition solutions can be achieved. The subjects of this study are to test a variety of conditions for the digestion of biological samples and to determine the contents of trace elements using ICP-AES and ICP-MS.

2 EXPERIMENTAL

Instrumentation

ICP-MS. The VG PlasmaQuad 2+ (VG Elemental, Ltd., Winsford, Cheshire, England) running under standard conditions was used for the experiments with ICP-MS.

ICP-AES. A Labtest V-25 Vacuum Spectrometer (grating 2160 L/mm, optical path 1m, 35 exit slits) and a Labtam Sequential Monochromator were used for the experiments with ICP-AES. The analytical lines selected for this work are listed in Table 1.

Table 1 Emission lines used for ICP-AES measurements

Element	Wavelength	Element	Wavelength
Cu	324.75 nm	C	247.86 nm
P	178.28 nm	Be	313.11 nm
S	180.73 nm	Te	238.58 nm
Zn	213.86 nm		
Mg	279.81 nm		

Table 2 Materials for decomposition experiments

Milkpowder	NIST 1549
Citrus Leaves	NIST 1572
Fish Tissue	IAEA MA-B-3/TM
Bovine Liver	BCR 185
Margarine	(> 50 % unsaturated fatty acids)

Decomposition. A High Pressure Asher ® (HPA) [2] (available from Hans Kuerner Analysentechnik, Rosenheim, Germany) with 70 ml quartz vessels was used as a high temperature decomposition apparatus for all experiments.

Materials

The materials tested are representing the varieties of food analysed in our laboratory. As far as possible they contain Standard Reference Materials covering the range from plant material to animal tissue up to fatty materials (Table 2).

Reagents

Nitric acid (subboiling grade) was used throughout the experiments. The water was purified with a Millipure System.

Sample Treatment

Digestion. Sample weights were 0.2-0.3g for each type of material. 3 ml of nitric acid were added for decomposition in the HPA. By means of the HPA a complete digestion of organic samples can be achieved at more than 280 °C [3]. In order to test less efficient decomposition conditions the following 4 ashing temperatures were applied to the samples:

150 °C,   200 °C,   250 °C,   300 °C.

The heating program consisted of 30 min heating up, 60 min hold at the maximum temperature and then cooling down. For each sample 3 repeats were carried out at each

temperature to obtain the variations for the decomposition step.
After digestion the solutions were diluted with water up to a final volume of 20 ml.
For measurement the solutions were diluted with water 1:1 for ICP-AES and 1:5 for
ICP-MS.

3 RESULTS AND DISCUSSION

Residual Carbon

The amount of residual carbon characterises the completeness of the oxidation
procedure and represents the soluble organic residues in the decomposition solutions.
This parameter contains information about the interferences which might be expected.
The results for residual carbon are obtained by measuring the carbon line with ICP-
AES. Figure 1 summarises the results at different ashing temperatures for all materials
tested. High amounts of residual carbon could be detected at low ashing temperatures:
the results range from 4-30% depending on the material. In animal tissues about 2% of
the sample is not completely oxidised at 250 °C ashing temperature. If 300 °C were
applied during the decomposition, the residual carbon ranges below 0.5% for all materials investigated.

The quality of the digestion is mainly determined by the temperature of the ashing
procedure, not by the length of this step. The completeness of digestion is controlled by
the oxidation potential of the nitric acid, which is a function of the ashing temperature
applied[4] . The fact that longer times will not influence the results significantly, is demonstrated by the results for ICP-MS for ^{52}Cr in Bovine Liver.

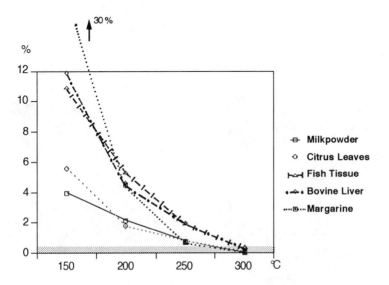

Figure 1 Residual carbon in the materials at different decomposition temperatures

ICP-AES

Results obtained with this method are presented in Figures 2a,b. The certified values of the SRM´s with their confidence interval are marked in shaded areas. The results for zinc and copper in Citrus Leaves (Figure 2a), phosphorus in Fish Tissue and magnesium in Milkpowder (Figure 2b) correspond very well with the certified values at each temperature. These examples indicate, that the qualitiy of the oxidation process has no influence on the results obtained.

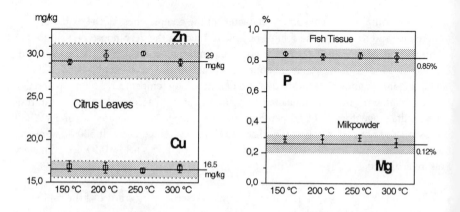

Figure 2   a) Results for Cu and Zn in Citrus Leaves
                b) Results for P in Fish Tissue and Mg in Milkpowder

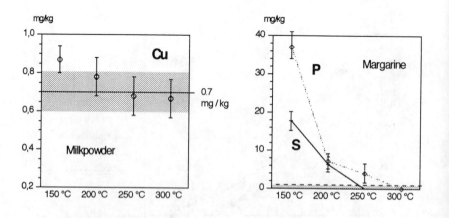

Figure 3   a) Results for Cu in Milkpowder and b) for P and S in Margarine

Figure 3a,b represents measurements near the detection limit of the ICP-AES method. The results for copper at the low ashing temperature indicate higher contents compared to the ashing temperatures 250 and 300 °C. They range outside the confidence interval of the certified value. For Margarine no information is available about the contents of sulphur and phosphorus. But the results obtained with ICP-AES decrease with increasing ashing temperature. Above 250 °C the values are below the detection limit of the method. These results are due to soluble organic residues in the solutions, which cause carbon bands in the emission spectra. The carbon lines are of low intensity but they may influence low level elemental determinations.

ICP-AES measurements were carried out with Be and Te as internal standards too. No nebulising effect depending on the ashing temperature can be detected.

ICP-MS

Element specific informations obtained with ICP-MS can be influenced by molecular interferences, thus restricting the specifity of the method[5,6]. One part of the molecular interferences can be caused by carbon-species e.g.: HCN affecting ^{27}Al, or ArC affecting ^{52}Cr. In samples containing molybdenum, MoC may influence the intensities of the silver isotopes 107 and 109.

Aluminum. Results for this element are presented in Table 4 . There is no information available from BCR for the content of Al in Bovine Liver. Own investigations with Graphite Furnace Atomic Absorption Spectrometry (GFAAS) produce a value of 2.50 ±0.20 mg/kg. The results obtained with ICP-MS range within this value, thus indicating that no influence on the Al-isotope depending on the decomposition temperature can be detected.

Chromium. Results for Fish Tissue and Bovine Liver, Margarine and Milkpowder are displayed in Figure 4. They indicate that the decomposition temperature has a significant influence on the chromium values. In Fish Tissue, the results at an ashing temperature of 150 °C are nearly double the certified value of 0.64 mg/kg. If the decomposition temperature is increased, the chromium content reaches the confidence range of the certified value and corresponds very well with this value at a temperature above 240 °C.

Table 4 Results for Al in Bovine Liver at different decomposition temperatures

Temperature	Al mg/kg	±
150 °C	2.56	0.12
200 °C	2.40	0.08
250 °C	2.80	0.25
300 °C	2.70	0.13

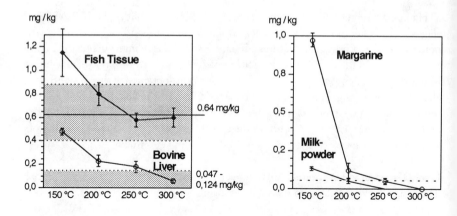

Figure 4 Cr-results a) in Fish Tissue and Bovine Liver b) in Margarine and Milkpowder

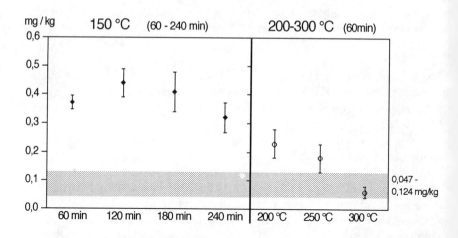

Figure 5 Influence of time and temperature on the results of ^{52}Cr in Bovine Liver.

The Cr-content in Bovine Liver is not certified. BCR reports an indicative value, ranging from 0.047 - 0.124 mg/kg, due to inhomogeneous material. The results obtained with ICP-MS show the same tendency as discussed before: the values for Cr are high at low decomposition temperatures and decrease when the ashing temperature is increased. This material needs temperatures higher than 250 °C, in order to achieve results that range within the indicative value.

The results cannot be enhanced by applying longer hold times for the ashing step (Figure 5). No significant decrease in carbon interferences can be achieved by longer hold times at 150 °C, compared to higher temperatures for the ashing step.

Milkpowder and Margarine are not certified for chromium, but the high values for this element at 150 °C indicate an interference caused by ArC on ^{52}Cr (Figure 4b). The results for chromium are below the detection limit, if ashing temperatures of more than 240 °C are applied.

Silver   Interferences at the silver isotopes 107 and 109 can be tested with Bovine Liver. This material contains remarkable amounts of about 2 mg/kg for molybdenum. The results for silver obtained with ICP-MS, compared to GFAAS, indicate no significant differences between the two methods within the ashing temperatures applied. At this level of molybdenum no interference on the silver isotope, caused by soluble organic compounds can be detected. The variations for ICP-MS are significantly higher than those obtained with GFAAS, due to the 1:5 dilution of the solutions for ICP-MS measurements.

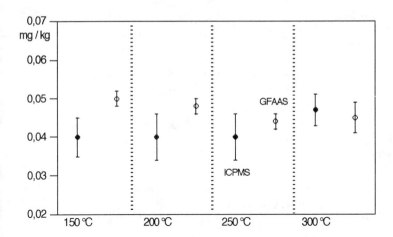

Figure 6   Results for silver in Bovine Liver with ICP-MS and GFAAS at
different ashing temperatures

4 CONCLUSIONS

The quality of the digestion solution depends on the type of organic material minera-lised and the temperature applied in the decomposition step. The content of the residual carbon in the solution represents the quality of digestion and corresponds directly to interferences in ICP-AES and ICP-MS.

Interferences in ICP-AES can be detected only if low levels of elements are deter-mined. The influence depends on the element determined and is due to carbon bands in the emission spectra. If the decomposition temperature is set to more than 240 °C, no influence can be detected for the materials investigated.

In ICP-MS severe interferences were found on ^{52}Cr caused by ArC. The results for this isotope are significantly influenced by the decomposition temperature and corres-pond directly to the soluble organic residues in the solutions. Difficult materials like liver or fish tissue need ashing temperatures of 280 - 300 °C for complete destruction of the organic matter and to achieve results, that are due to the real Cr-content in the samples.

5 REFERENCES

1 R. Bock, `A Handbook of Decomposition Methods in Analytical Chemistry´, International Textbook Company, London, 1979.
2 G. Knapp , ICP Inf. Newsl., !986, 12, 273.
3 M. Würfels, E. Jackwerth and M. Stoeppler, Fresenius´ Z. Anal. Chem., 1988, 330, 159.
4 M. Würfels and E. Jackwerth, Fresenius´ Z. Anal. Chem., 1985, 322, 354.
5 A.R. Date and A.L. Grey, `Applications of Inductively Coupled Plasma Mass Spectrometry´, pp 29-32, Blackie, London, 1989.
6 S. Yamasaki, A.Tsumura and D.Cai, in G.Holland and A.E.Eaton `Applications of Plasma Source Mass Spectrometry´, pp 110-118, The Royal Society of Chemistry, Cambridge, 1991.

A Dry Chlorination/Microwave Digestion/ICP-MS Analytical Method for the Determination of the Platinum Group Elements and Gold in Metallic and Non-metallic Fractions of Rocks

B. J. Perry, R. R. Barefoot, J. C. Van Loon, and
A. J. Naldrett
DEPARTMENT OF GEOLOGY, UNIVERSITY OF TORONTO,
CANADA M5S 3B1

D. V. Speller
INCO LTD., J. ROY GORDON RESEARCH LABORATORY, SHERIDAN
PARK, MISSISSAUGA, ONTARIO, CANADA L5K 1Z9

1   INTRODUCTION

This paper describes a new analytical method for the determination of most platinum group elements (PGE's) and Au in two fractions of rocks: the metallics fraction (native metals, natural alloys and sulphide group minerals); and the non-metallics fraction (silicate and refractory oxide minerals). PGE's (except Os, Ru) and Au contained in the metallics fraction are determined by a dry chlorination / inductively coupled plasma-mass spectrometry (ICP-MS) method[1]. PGE's (except Os, Ru) and Au contained in the non-metallics fraction (the remainder after dry chlorination of rock pulp), are determined by a microwave digestion / ICP-MS method developed by the authors and first described herein. While the microwave digestion /ICP-MS method presented in this paper could be improved with minor refinements, it can be used for expeditious quantitation of PGE's and Au in the non-metallics fraction of rocks. In addition to information regarding the residence of PGE's and Au in both the metallics fraction and the non-metallics fraction, the two methods combined provide the determination of total PGE's and Au in the rock sample.

The purpose of the development of the microwave digestion / ICP-MS method for the determination of PGE's and Au in the non-metallics fraction was to provide a means of investigating the detectability of PGE depletion in layered mafic intrusions that host PGE deposits. PGE deposits in layered mafic intrusions result from a three-step process[2]: NiS droplet formation in the magma; collection of PGE's from the magma by the settling NiS droplets; and the accumulation of PGE bearing NiS droplets on or near the floor of the magma chamber. Thus, PGE depletion of the magma is a prerequisite to the formation of this type of PGE deposit. The ability to detect PGE depletion has great potential benefit to the mineral exploration industry. The large volume of PGE-depleted magma required to produce an economic PGE deposit should present a much larger exploration target (many square kms) than does the narrow horizon into which the NiS droplets have settled, usually less than a few meters thick. Data resultant from the application of this method to Fox River Sill samples (Manitoba, Canada) are presented to show the analytical suitability of this method for the detection of PGE depletion.

In order to determine PGE's resident in the non-metallics fraction of the rock pulp, the metallics fraction must be removed first. The removal of the metallics fraction cannot be accomplished quantitatively by digestions relying on either simple or mixed acids because of the well known variable reactivity of noble metals towards mineral acids, as previously elucidated by Beamish and co-workers[3]. Furthermore, many naturally occurring PGE alloys (e.g. iridosmine, platiniridium) are exceptionally resistant to acid attack, even by hot *aqua regia*. Recently, Gowing and Potts[4] evaluated an *aqua regia* leach as a collection step for PGE's in rock pulps. Pd and Au were acceptably recovered. Only 10 to 40% of Pt, Rh, Ru, Ir, and Os were recovered.

However, chlorine at about 600°C attacks all the noble metals[5]. After extensive studies of the effects of various chlorination methods (direct, wet and dry) on PGE's and Au as pure metals and alloys, Beamish[3] strongly recommended chlorinations for the quantitative conversion of PGE's, Au and their alloys into complex salts. Dry chlorination, accomplished by passing hot chlorine gas over the sample either alone or mixed with a small amount of NaCl in an open tube at 500-600°C, produced PGE and Au bearing complex salts. The salts, produced in the presence of admixed NaCl, are soluble in weak HCl. Rocks were not chlorinated in these investigations. Van Loon and co-workers[6] applied dry chlorination in an open tube system to sulphide concentrate, Cu bullion alloy, precious metal coated glass wool, and Fe-Ni concentrate. In each case they obtained acceptable agreement with fire assay results. Perry, Speller and Van Loon[1] developed and tested a dry chlorination /ICP-MS method for the determination of PGE's (except Os) and Au in rock pulps. They obtained quantitative results for certified reference rock pulps (SARM-7, Mintek, South Africa; NBM 6a and 6b, Nevada Bureau of Mines, U.S.A), an INCO Ltd. in-house PGE and Au reference pulp (S-1800), and ordinary gabbro rock. It is this method that the present authors use to remove the metallics fraction before determining PGE's and Au in the remaining non-metallics fraction by microwave digestion and ICP-MS analysis of the digestate.

Although microwave heating in wet ashing procedures was known as early as 1975, it was not until 1983 that it was first applied to geological materials in closed vessels[7]. These investigators used HF-HNO$_3$-HCl acid mixtures in closed polycarbonate bottles to dissolve steel slag, feldspar and Ni-Cu alloy. The higher temperatures and pressures required to dissolve more intractable geological species (zircons, ores) were later obtained by other investigators by using closed Teflon PFA vessels, from which the currently available commercial microwave acid digestion vessels have evolved. Recently Totland, Jarvis and Jarvis[8] assessed dissolution techniques for the analysis of geological samples by plasma source spectrochemistry. They cite low quantitation limits, very low blanks, better dissolution of refractory minerals, rapidity and low cost of consumables as advantages of microwave digestion. Unsuitability for the determination of some elements and high capital cost (of commercial analytical microwave ovens and digestion vessels) are indicated as disadvantages.

2   EXPERIMENTAL

Equipment and materials

Chlorination tube (manufactured at the University of Toronto). An all Vycor chlorination tube with an expanded chlorination chamber (2.5 cm x 45 cm, 85 cc) contains interior longitudinal ribs which help provide efficient contact between the rock pulp, admixed NaCl and slowly flowing Cl$_2$ gas. At both ends of the chlorination chamber there is a sharp taper to 10 mm ID Vycor tube (10 cm entrance section, 15 cm condensation section). The outside end of the entrance section of the chlorination tube is connected by 1.4m of Tygon tubing to a chlorine gas regulator (Matheson Gas Products Canada Inc., Toronto). The regulator is connected to an Argon purging valve (Matheson), which is connected to a 31 kg cylinder of high purity Cl$_2$ gas (Matheson). The regulator is purged with Ar after each use. The exit portion of the chlorination tube is connected to a fritted glass bubbler submerged in 1L of water, via 1.4m of Tygon tubing. After bubbling through 25 cm of water, the fumes are vented directly into the fumehood draft vent via Tygon tubing.

Tube furnace. The chlorination chamber portion of the chlorination tube was entirely contained within a split shell tube furnace (600W, Heavy Duty Electric Co., Watertown, WI, U.S.A.). A Variac autotransformer controls power to the resistive heating elements of the furnace. Aluminum support plates are attached outside of the furnace at both ends in order to keep the tube centered and supported while it is rotated/agitated. Woven fiberglass insulation (2 cm wide) is wrapped around the portion of the tube that resides in the space between the support plates and the sides of the furnace. This was done to help keep the temperature in the chlorination chamber more uniform throughout its length, to provide a more abrupt temperature drop after the chlorine gas leaves the chamber and to facilitate the condensation of volatile PGE chlorides onto the glass wool plugs. Temperatures within the furnace chamber are measured before and after the chlorination with a digital thermometer (model HH-51, Omega Electronic Accessories, Toronto, Canada). The thermocouple, enclosed in a Vycor tube similar to the chlorination tube, is inserted into the furnace chamber when a temperature measurement is required.

Glass wool plugs. Borosilicate glass wool (Fisher Scientific Ltd., Toronto, Canada) is used to make 4 small plugs (1 cm x 2 cm) that are inserted just outside the furnace at the junctions of the narrow diameter entrance/exit tubes and the chlorination chamber and at the junctions of the Tygon tube and the chlorination tube. Besides retaining the rock pulp in the chamber, the two innermost plugs collect the condensate of PGE chlorides which are volatile.

Microwave oven. Toshiba, ERX-1610, household appliance.

Chlorine. High purity chlorine gas (Matheson).

Sodium chloride. Powdered, high purity NaCl (AnalaR, BDH).

Filtration. Millipore, vacuum unit; 48 mm, 0.45 um pore size HA filters.

Hydrofluoric acid. 48-51%, trace metal grade (Fisher).

Hydrochloric acid. 37-38%, ACS reagent grade (Fisher).

Nitric acid. 69-71%, ACS reagent grade (BDH).

Microwave acid digestion bombs. Parr #4781, Teflon cup, 23 ml.

Inductively Coupled Plasma - Mass Spectrometer. A SCIEX ELAN 250 ICP-MS, with ion optics upgraded to ELAN 500, was operated at 1.2 kW forward power with <5 watts reflected power. Torch gas flows were: plasma, 12.0 l min.$^{-1}$; nebulizer 38 psi; auxiliary 1.4 l min.$^{-1}$. A peristaltic pump provides 0.9 ml min.$^{-1}$ solution uptake to a Meinhard concentric nebulizer mounted in a Scott type spray chamber.

Sample descriptions and preparations

The rock pulps were supplied by the sources indicated below. The pulps were not re-ground by the present authors.

INCO (0.1)S-1800. This material is an INCO Ltd. in-house precious metals reference material composed of Frood-Stobie Mine (Ni, Cu, PGE's) tailings, predominantly noritic rock containing small amounts of

PGE bearing Ni and/or Cu sulphide minerals. This material passes through Tyler series 100 mesh.

Fox River Sill. These mafic-ultramafic rock samples were obtained from Dr. R. F. J. Scoates (Geological Survey of Canada) as splits from a length of diamond drill core representing a vertical section from the surface through at least one PGE mineralized layer of the Fox River Sill and continuing also below the mineralized section. The splits were ground to -180 mesh by XRAL Laboratories, Toronto, Canada.

3   METHOD

Dry-chlorination

One glass wool plug is inserted in the condensation section of the dry-chlorination tube just after the chlorination chamber, and a second glass wool plug is inserted at the end of the tube. NaCl is added to each sample (0.1g/15g rock pulp) by mixing the appropriate amount of powdered NaCl into the rock pulp, placing the admixed NaCl and rock pulp in a 125 ml polypropylene bottle and adding 20 ml distilled de-ionized water (ddw). The bottle and contents are shaken by hand for 15 s and then allowed to stand uncovered in a drying oven pre-heated to 80°C until the pulp is dry. The bottle is re-capped and shaken by hand. The contents are emptied out onto 15 cm x15 cm glassine weighing paper and then are transferred to the chlorination chamber through the entrance section.   A third glass wool plug is inserted through the entrance section of the chlorination tube and is positioned at the junction of the entrance tube and the beginning of the chlorination chamber. A fourth glass wool plug is positioned at the outside end of the entrance section of the chlorination tube.   In the fumehood, the chlorination tube containing the sample is connected to the chlorine gas line and to the bubbler tower line.   Air in the chlorination tube is displaced with chlorine. The chlorine flow is reduced to only a few bubbles per s. The slow flow of chlorine is maintained throughout the chlorination. With chlorine flowing through, the tube and its contents are heated for 3.5 h in a tube furnace pre-heated to 580° C. The tube and its contents are briefly agitated by hand every 20 min. After cooling while chlorine continues to flow, the chlorination products and the glass wool plugs are placed in a 250 ml Pyrex beaker. Three rinses of the emptied tube (50 ml total, warm 10% HCl) are also added to the beaker. The watch glass covered beaker and its contents are heated to 80° C for 10 min. The contents are vacuum filtered through a 0.45 μm MSI cellulose nitrate membrane filter, directly into a culture tube pre-marked at 50 ml. The filtrate is brought to volume in the culture tube by addition of ddw. Two ml are pipetted to a 10 ml volumetric flask and diluted 1:5 with ddw. The diluted solution is analyzed for PGE's and Au by ICP-MS.

Microwave digestion

The chlorination resistant (non-metallics) residual fraction remaining on the filter is rinsed with 50 ml ddw by adding this volume of ddw to the funnel containing the chlorination resistant fraction and by applying vacuum to draw the ddw through the solids. The rinse is repeated. The chlorination resistant fraction is removed from the filtration unit and is transferred to a glassine weighing paper. It is then allowed to dry for 15 min. in an oven pre-heated to 80°C. Glass wool fibers are separated from the dried sample by hand sieving (-100 mesh). Two grams are placed into the Teflon cup of the Parr microwave acid digestion bomb. Two ml of HNO_3 and 10 ml HF are added to the Teflon cup containing the sample. The cup is placed in the bomb, and the bomb is sealed. Microwave heating is applied at high power for three 1 min. heating periods. Between each heating period, the bomb is allowed to cool in the draft of the fumehood for 5 min. The bomb is allowed to cool for 30 min. in the draft of the fumehood before it is opened. After the final cooling period, the bomb is opened and 1 ml HCl is added to the contents of the Teflon cup. The cup is recapped and shaken briefly by hand. The digestate is poured into a 30 ml Nalgene high density polypropylene bottle. One ml is removed and diluted 1:25 with ddw in a 25 ml volumetric flask. The diluted digestate is analyzed by ICP-MS for PGE's and Au.

Instrumental analysis

The mass spectrometer is programmed to detect ^{102}Ru, ^{103}Rh, ^{106}Pd, ^{192}Os, ^{193}Ir, ^{195}Pt, ^{197}Au. Sequential peak hopping is employed. Dwell is 50 ms with ten repeats, before hopping to the next peak. The peak hopping sequence is performed six times. Each sample solution is analyzed by the standards addition method. Spiked sample and blank solutions (+10 ng g^{-1} PGE's and Au) are prepared immediately preceding instrumental analysis. Ten μl of 1 μg g^{-1} mixed PGE's and Au standard are added to 1 ml of the sample solution. The sample solution is analyzed, followed by a 30 s rinse out with 1% HNO_3. The 1% HNO_3 is then analyzed in order to detect carry over. Blank intensities are usually achieved after only one rinse out. The spiked sample is then analyzed. This is again followed by a 30 s rinse with 1% HNO_3, then analysis of the 1% HNO_3. Quantitation (ng g^{-1}) is achieved by: deducting the sample counts per second (cps) from spiked sample cps to yield the net cps for 10 ng g^{-1} for each element; dividing the sample cps by the net cps for 10 ng g^{-1}; multiplying by 10 and by the appropriate dilution correction factors. The concentrations of PGE's and Au in the blanks are also obtained in this manner. The results for the blanks are deducted from the results for the sample solution in order to obtain the concentrations of PGE's and Au in the sample pulp.

4   RESULTS AND DISCUSSION

The present authors have made several minor modifications to the dry chlorination method as originally proposed. Previous work[1] showed that the presence of NaCl was essential to obtaining high, reproducible PGE and Au recoveries from reference rock pulps. Beamish[2] noted that in the absence of admixed NaCl, some of the PGE chlorides produced by the chlorination process were insoluble in water and acids. He also found that in the absence of admixed NaCl some of the PGE chlorides (especially Ru, Os chlorides) produced by the chlorination process were volatile at temperatures encountered in the chlorination tube. This could result in losses of PGE's as these volatile chlorides are carried out of an open tube by the flowing Cl_2 gas. It is likely that the presence of NaCl promotes the production of non-volatile sodium chloroplatinates, chloropalladinates, chlororhodates, chloroiridates and chloroaurates, each of which are soluble in weak HCl. The present authors use less NaCl (0.1g/15g sample vs. 0.5g/15g sample), but this smaller amount is more efficiently distributed through out the sample by dissolving it in a ddw/rock pulp slurry and drying the slurry before chlorination. This results in a 5 times reduction in NaCl in the final solution (0.2% vs. 1.0%) and preserves quantitative recovery of PGE's and Au (Table 1). Reducing the NaCl level to 0.05g NaCl resulted in diminished recoveries.

TABLE 1.   Recovery of PGE's and Au (ug g^{-1}) from (0.1)S-1800 as a function of introduced NaCl.

NaCl/15g pulp		Pt	Pd	Rh	Ir	Au
0.25g admixed	n=4	18	171	7.6	1.2	13
0.10g admixed	n=3	21	71	7.7	1.2	13
0.10g dissolved	n=4	24	155	8.8	1.3	11
0.05g dissolved	n=3	13	51	3.5	0.7	10

TABLE 2.   Suppression(%) of net signal intensity for PGE's and Au in (0.1)S-1800 as a function of introduced NaCl.

NaCl/15g pulp		Pt	Pd	Rh	Ir	Au
0.5g	n=2	70	80	88	88	79
0.25g	n=4	19	42	63	<10	<10
0.10g	n=3	<10	14	<10	<10	<10

% suppression = {1 - [net intensity/(ddw+10 µg g^{-1})]} x 100
net intensity = cps spiked sample(+10 µg g^{-1}) minus cps sample.

The lower Na content used by the present authors also reduces the adverse effects of Na-induced ionization suppression during ICP-MS analysis of the final solution (Table 2). For each sample the final volume resultant from the chlorination procedure is 50 ml.

Previously, signal intensity in the mass spectrometer was diminished continuously because of gradual build-up of salts on the skimmer and sampler cones. As the salts form, the sampler orifice effective diameter is gradually reduced, the amount of plasma sampled decreases and the analyte signals diminish. In addition to reducing the amount of added NaCl, the present authors dilute (with ddw) the solutions resultant from dry chlorinations by factors of 5 or 10, and those from microwave digestions by factors of 25 or 50. This nearly eliminates salt build up on the sampler and skimmer orifices. Ionization suppression due to large amounts of dissolved solids in the undiluted solutions is very much reduced, especially in the case of solutions resultant from the microwave digestion procedure (Table 3).

TABLE 3.   Suppression(%) of net signal intensity for PGE's and Au in (0.1)S-1800 as a function of dilution.

sample/dilution		Pt	Pd	Rh	Ir	Au
M95-4/10	n=2	85	90	85	81	90
M95-4/25	n=3	34	24	27	<10	41
M95-4/50	n=3	13	15	<10	<10	17

% suppression = {1 - [net intensity/(ddw+10 ug g^{-1})]} x 100

net intensity = [(cps sample spiked+10 ug g^{-1}) - cps sample].

Blanks

As a safeguard against PGE contamination within the Cl_2 gas or as picked up by the Cl_2 gas from the regulator, we have installed a graphite trap between the regulator and the chlorination tube This consists of 15 cm of 1 cm ID Tygon tubing packed with small chips of coconut shell activated charcoal held in place within the tubing by 3 cm glass wool plugs.

Blanks carried for the dry chlorination method consisted of NaCl, glass wool plugs and Cl_2 gas. Blanks carried for the microwave digestion consisted of HF, HNO_3 and HCl. PGE and Au contaminations in the blank solutions were low, except for Pd and Au in the microwave digestion blank (Table 4).

TABLE 4. Concentrations of PGE's and Au (ng g^{-1}) in blanks

		Pt	Pd	Rh	Ru	Ir	Au
chlorination (20g)	n=3	0.1	0.8	0.1	0.3	0.04	0.3
microwave (2g)	n=3	2.1	37	0.8	3.0	1.2	5.8

At the low PGE and Au concentrations most often encountered in exploration rock samples (<50 ng g^{-1}) analytical precision, both within-solution and within-sample, is often surprisingly good (Cv's <10%). However, our experience is that random fluctuations occur. Thus, triplicate analyses of each sample, and each solution, is recommended.

Fox River Sill samples

The Fox River Sill samples originate from a PGE mineralized layer and the non-mineralized layer immediately above it. The analytical method applied to these samples provides PGE and Au residence data that can be used to make observations both within-layer and within-fraction. These observations can then be compared between layers and between fractions.

Within the non-mineralized layer (samples #138, #142) PGE and Au concentrations are much higher in the non-metallics fraction than in the metallics fraction (Table 5). The opposite situation exists within the mineralized layer (samples #220, #225, #236), where PGE and Au concentrations are usually lower in the non-metallics fraction than in the metallics fraction. This difference can be seen most clearly by comparing PGE and Au residence ratios, i.e. (non-metallics fraction PGE's and Au) / (metallics-fraction PGE's and Au), between the two layers (Table 6). For the non-mineralized layer, this ratio is much greater than one for each PGE and for Au. However, for the mineralized layer, this ratio is less than one for each PGE and for Au.

TABLE 5.   Concentrations of PGE's and Au (ng g^{-1}) in Fox River Sill (ddh#2) samples after dry chlorination and after microwave digestion.

		Pt	Pd	Rh	Ir	Au
non-metallics fraction						
#138	n=2	31	208	17.4	19	56
#142	n=3	28	205	12.0	15	38
#220	n=3	12	nd	2.1	nd	25
#225	n=3	16	34	4.7	0.9	41
metallics fraction						
#138	n=3	1.2	nd	2.1	0.7	7.1
#142	n=3	2.5	5.9	2.0	1.2	7.3
#220	n=3	28	21	7.3	9.9	42
#225	n=3	25	32	3.8	11.8	44
#236	n=3	26	36	5.1	12	45

nd=not detected

TABLE 6.   PGE and Au residence ratios within a mineralized layer and within a non-mineralized layer of the Fox River Sill.

	Pt	Pd	Rh	Ir	Au
non-mineralized layer	16	69	14	18	6.5
mineralized layer	0.5	0.6	0.6	0.04	0.8

(non-metallics fraction PGE's and Au) / (metallics-fraction PGE's and Au)

Within the non-metallics fraction, PGE and Au concentrations are higher in the non-mineralized layer than in the mineralized layer. Within the metallics fraction, PGE and Au concentrations are higher in the mineralized layer than in the non-mineralized layer. These two observations, together with the observed shift in PGE and Au residence ratios between the two layers, suggest that in the mineralized layer a process may have occurred that depleted the non-metallics fraction of its PGE's and Au, and enriched the metallics fraction in PGE's and Au.

5   ANALYTICAL RESEARCH IN PROGRESS

More testing of the microwave digestion portion of this method is required, especially on samples of rocks from ore environments that contain predominantly non-metallic PGE ore minerals, e.g. chromite, magnetite. In addition to further testing on different rock types, we are presently investigating the suitability of the two methods for the determination of base metals (especially Ni, Cu, Zn and Co) and traditional geochemical exploration pathfinder elements (especially Te, Sb and As).

6   CONCLUSIONS

A method comprised of dry chlorination and $HF-HNO_3-HCl$ microwave digestion, followed by ICP-MS analyses of the resultant solutions, has been used to determine low concentrations of PGE's and Au in two fractions of rocks (metallics fraction and non-metallics fraction). The data resultant from the application of the method suggests that the combined method has good analytical potential for use in investigations concerned with the detection of PGE depletion in magmas as a potential exploration indicator of the presence of PGE bearing NiS deposits.

7   ACKNOWLEDGMENTS

The authors thank INCO Exploration and Technical Services (I.E.T.S.) for the generous funding that has made this work possible. In particular we are grateful for the support, encouragement and guidance of Dr. R. A. Alcock, Manager, Geological Research and Laboratories, INCO Ltd./I.E.T.S. We gratefully acknowledge Perkin-Elmer SCIEX for contributing important upgrades to the ELAN 250 ICP-MS. In particular, Ken Halligan (SCIEX) most generously shared his great technical expertise. This research was also supported by the Natural Sciences and Engineering Research Council of Canada through an Operating Grant (Van Loon) and a Strategic Grant (Naldrett and Van Loon).

REFERENCES

1. Perry, B. J., Speller, D. V. and Van Loon, J. C., A Dry-Chlorination Inductively Coupled Plasma Mass Spectrometric Method for the Determination of Platinum Group Elements and Gold in Rocks, *JAAS*, 1992, September, vol 7, in press.

2. Naldrett, A. J. *Magmatic Sulphide Deposits.* Oxford Monographs in Geology and Geophysics, Clarendon Press, New York. Oxford University Press, Oxford. 185pp.

3. Beamish, F. E., *The Analytical Chemistry of the Noble Metals,* Pergammon Press, Oxford, 1966, 608pp.

4. Gowing, C. J. and Potts, P. J. Evaluation of a Rapid Technique for the Determination of Precious Metals in Geological Samples Based on Selective *Aqua Regia* Leach. *Analyst,* 1991, vol. 116, pp. 773-779.

5. Van Loon, J. C. and Barefoot, R. R., *Determination of the Precious Metals: Selected Instrumental Methods,* Wiley, Chichester, England. 1991, 276pp.

6. Van Loon, J. C., Szeto, M., Howson, W. W., and Levin, I.A., A Chlorination Atomic Spectrometry Method for the Analysis of Precious Metal Samples *At. Spectrosc.,* 1984, **2**, 43.

7. Farrell, R. F., Matthes, S. A. and Mackie, A. J. A Microwave System for the Acid Dissolution of Metal and Mineral Samples. U. S. Bureau of Mines Analytical Support Services Program, Technical Progress Report 120, 1983, 9pp.

8. Totland, M., Jarvis, I. and Jarvis, K. E., An Assessment of Dissolution Techniques for the Analysis of Geological Samples by Plasma Spectrometry. In: I. Jarvis and K. E. Jarvis (Guest-Editors), Plasma Spectrometry in the Earth Sciences. Chem. Geol., 1992, 95: 35-62.

Evaluation of an Ion Exchange Separation of Cs, Sr, and Y by ICP-MS

E. Vega-Rangel, F. Abou-Shakra, and N. I. Ward

DEPARTMENT OF CHEMISTRY, UNIVERSITY OF SURREY, GUILDFORD, SURREY GU2 5XH, UK

1 INTRODUCTION

Nuclear reactor accidents and the continuous testing of nuclear weapons have resulted in increasing distribution of radioactive fission products in the environment. Some radionuclides such as ^{90}Sr and ^{137}Cs are extremely hazardous, from the point of view of environmental contamination, due to their long physical and biological life, ^{137}Cs ($t_{1/2}=30.17$y), ^{90}Sr ($t_{1/2}=28.5$y)[1]. Strontium enters the human diet via a complex series of routes in association with calcium and hence accumulates in bone tissue[2]. Similarly, the caesium isotopes are easily incorporated into the food chain and become uniformly distributed in the body tissue following competition with potassium[2]. Owing to their low levels, Sr and Cs are difficult to determine in environmental samples. Therefore, it is essential to undertake a preconcentration step before the analysis is carried out. Furthermore, it is necessary to separate Sr from its radioactive daughter product Y since both ^{90}Sr and ^{90}Y isotopes are pure ß emitters.

Several researchers[3,4,5] have described methods for the radiochemical analysis of high levels of ^{137}Cs and ^{90}Sr. However, most of these techniques are not suitable for dealing with the low levels normally encountered in environmental samples.

2 EXPERIMENTAL

Inductively coupled plasma mass spectrometry was used for the determination of Cs, Sr and Y. First, a series of preliminary experiments were carried out to determine the optimum conditions of the column based on recovery and peak resolution. Secondly, the retention behaviour of the cation exchange resin, used for separating synthetic solutions was evaluated. In particular, the influence of the molarity of the mobile phase, HCl (range 1 to 6 M), upon the degree of elemental separation was studied. Finally, ammonium molybdophosphate (AMP) was used in an attempt to improve the recovery of Cs.

Preparation of the Columns

A glass column (i.d. of 10 mm and 20 cm length) was packed (2.5 cm depth) with Bio-Rad AG 50W-X8 200-400 mesh hydrogen form (Bio-Rad Lab. Richmond, England). It was then rinsed with 50 ml deionised water (18 MΩ cm^{-1}), cleaned with 50 ml of HCl 6M (Analytical Reagent, Fisons Lab. England) and primed with 15 ml of 2.5 M HCl.

Optimum Conditions for the Column

In order to select the optimum conditions for the column the following studies were carried out.

Depth of the Resin. Columns were packed with different depths of resin (1.3, 2.6, 3.9, 5.2 and 7.8 cm), then rinsed with 50 ml of deionised water, finally 10 ml of acid were eluted by using a peristaltic pump and the time of elution was recorded. The results were 3.06, 5.37, 8.18, 10.55 and 16.01 min respectively. The recoveries were calculated for the different depths, by eluting 1ml of Sr at a concentration of 1μg ml^{-1} and 50 ml of acid as a eluant (4M HCl), the following recoveries being observed: 48.12, 52.9, 54.4, 55.8 and 56.5% respectively. From the results it can be seen that the best recovery was obtained when 7.8 cm of resin were used, but it is time consuming. Therefore it was decided to use 2.6 cm of resin, since it is a compromise between time (which is important when handling radioactive materials) and recovery.

Flow Rate. The flow rate was studied by varying the speed of the pump (Figure 1). As can be seen from the results, when higher speeds are used the area under the peak is wider, resulting in poor resolution of the peak. It was decided to use 1.5 ml min^{-1} since this gives the best peak resolution.

Gradient Elution and Recovery

The evaluation of the effect of various acid concentrations was studied. A synthetic solution containing 1μg ml^{-1} of Cs, Sr and Y was prepared using 1000 μg ml^{-1} standard solutions (BDH, Spectrosol Limited, England). A 1 ml aliquot of this solution was placed onto the column, eluted with 50 ml of 1M HCl, and a series of ten 5 ml samples were collected. These aliquots were then analysed to determine the elution peaks for the various elements (Fig. 2, 3 and 4). This procedure was repeated using different concentrations of HCl, namely 1.5, 4 and 6M.

The recoveries were determined by measuring the concentration after elution of 4 ng ml^{-1} of the synthetic solution. Standard addition was used to evaluate the accuracy of the method. From the eluate, six fractions of 1 ml each were respectively spiked with 0, 0.5, 1, 1.5, 2 and 2.5 ml of a standard solution containing 10 μg ml^{-1} of Cs, Sr and Y. For each sample 0.5 ml of In (100 ng ml^{-1}) were added as an internal standard. The solutions were made up to 5 ml with deionised water and analysed by ICP-MS.

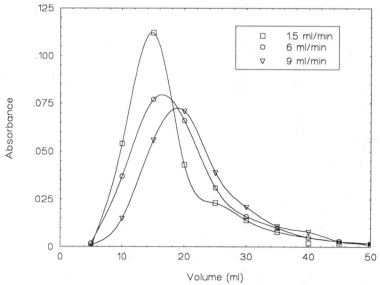

Fig. 1. Elution Curves for Sr using 4M HCl with different flow rates

AMP Type Resin

The recovery of Cs with gradient elution was 65%. Thus, to improve this figure a separation step was adopted. In this step AMP was added to the resin so that the solution has to first pass through this column before gradient elution. A 0.1 g portion of AMP was mixed with 2 g of the resin, 1 ml of Cs (1 μg ml^{-1}) was added and left overnight. The final solution was eluted with 50 ml of 1M NH$_4$OH. The method of standard addition was used to check the accuracy of the results.

Instrumental Operating Conditions

Stable isotope analysis (^{133}Cs, ^{88}Sr and ^{89}Y) was performed using the Finnigan MAT SOLA ICP-MS instrument (Finnigan MAT, England) in the Department of Chemistry, University of Surrey. The operating parameters used were: plasma incident power 1.5 kW, reflected power <5 watt, cooling gas flow rate 16 l min^{-1}, intermediate gas flow rate 1 l min^{-1}, nebuliser gas flow rate l l min^{-1} and a sample flow rate of 0.8 ml min^{-1}. Indium, at a concentration of 100 ng ml^{-1}, was added as an internal standard to all samples.

3 RESULTS AND DISCUSSION

Gradient Elution

Fig. 2 shows that, using 1M HCl, Cs can be separated from the other two elements. The observed recovery for Cs was 65%. After separating Cs, 1.5 M HCl can be used to elute Sr while Y is still retained on the column (Fig. 3).

Finally as shown in Fig. 4, 6 M HCl can be used to elute Y of the column. Thus gradient elution can be employed to separate all three elements over a 30 minute period.

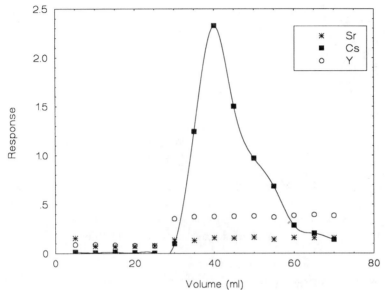

Fig. 2. Elution peaks using HCl 1M (Bio-Rad)

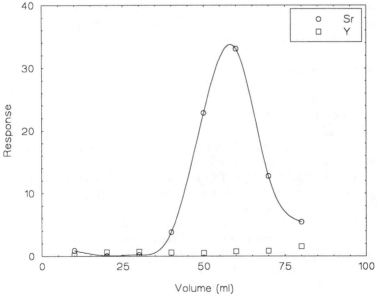

Fig. 3. Elution peaks using HCl 1.5 M (Bio-Rad)

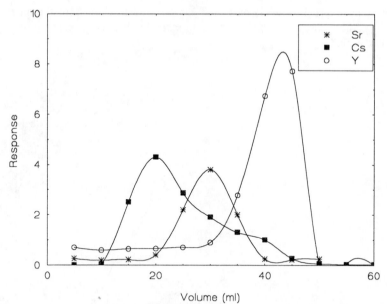

Fig. 4. Elution peaks using HCl 6 M (Bio-Rad)

Recovery of Cs Using AMP

The concentration of the eluate was calculated by using both external calibration and standard addition. No significant differences were observed between the two methods. The calculated recovery for Cs was 93%. This result shows that the use of the Bio-Rad in conjunction with AMP significantly improved the recovery of Cs. Two other bases, NaOH and NH_4Cl could be used as eluants, but they were ignored in this study because of the known effect of Na and Cl on the signal when using ICP-MS.

4 CONCLUSION

From the results it can be seen that a simple gradient elution can be used for the preconcentration of Cs and separation of Sr from its daughter Y in a relatively short time. Recovery of Cs by using the AMP in conjunction with the cation exchange resin Bio-Rad was improved to greater than 90%. This preliminary study has shown that the elution of Cs, Sr and Y using HCl on a Bio-Rad AG 50W-X8 ion exchange column is influenced by the molarity of HCl. Future work is now being undertaken to investigate the problems encountered with matrix effects in natural samples. It is anticipated that approved separation scheme will then be applied to the determination of ^{137}Cs and ^{90}Sr in environmental or dietary samples from Mexico.

Acknowledgment:
This work is being supported by a grant from the University of Mexico (UNAM).

References
1. HARVEY, B. R. et al. Aquatic Environment Protection: Analytical Methods. ISSN 0953-4466, Number 5, June 1989
2. EISENBUD, M., Environmental Radioactivity, 3rd edition. San Diego, CA. Academic Press, 1987.
3. E. BLASIUS and W. KLEIN, J. Radioanal. Nucl. Chem. Lett. ,1985, <u>126</u>, 389.
4. J. BORCHERDING and H. NIES, J. Radioanal. Nucl. Chem., 1986, <u>98</u>, 127.
5. K. BUNZL and W. KRACKE, J. Radioanal. Nucl. Chem., 1991, <u>148</u>, 115.

ICP-MS Determinations in Automotive Catalyst Exhaust

S. Knobloch and H. König
FRAUNHOFER INSTITUTE OF TOXICOLOGY AND AEROSOL RESEARCH,
NICOLAI-FUCHS-STR. I, D-W-3000 HANOVER 61, GERMANY

G. Wünsch
INSTITUTE OF INORGANIC CHEMISTRY, UNIVERSITY OF HANOVER,
CALLINSTR. 9A, D-W-3000 HANOVER I, GERMANY

1 INTRODUCTION

Automotive exhaust gases are one important source of air pollution. The use of three-way catalytic converters, containing noble metals such as platinum, lower the emissions of carbon monoxide, nitrogen oxide and hydrocarbons up to 90 percent [1]. To calculate the risk which may occur from platinum emissions, a working group called "Noble Metal Emissions" was founded by acknowledgement of the government. In our laboratory we test three-way monolithic converters from four manufacturers under varied conditions. First results of our study are outlined.

2 EXPERIMENTAL

Apparatus

A computer controlled dynamometer was used to apply different loads to a 1,8 l Passat engine (69 kW). The engine was equipped with different commercially available three-way catalytic converters. Two different driving conditions were simulated: Constant speed of 80 km/h; an urban driving cycle called US 75.

Sampling Procedure

1) Particle samples were taken semiisokinetically (US 75 Cycle) or isokinetically (constant speed 80 km/h) directly from the exhaust pipe of the engine behind the first silencer (figure 1). The particles were collected on teflon foils. They were classified using a five stage Berner impactor with backup filter with a nominal flow rate of 9 Nm^3 and a cut-off diameter between 0,12 and 10,2 μm. To avoid bounce-off effects, the deposition foils were impregnated with a silicon adhesive.

2) In order to estimate the volatile fraction of the automobile exhaust we used an apparatus developed by US EPA with special focus on sampling organic compounds [2]. It consists of a heated glassfibre filter to remove particulates, a glass-condenser to collect water and the fraction of the condensable gaseous compounds which we collect in a cooling trap.

Sample Preparation

1) The particle samples of each stage of the impactor were digested in a high pressure asher with nitric acid and hydrochloric acid in the ratio of 3 to 1, at a maximum temperature of 240°C and a nitrogen pressure of 100 bar. These solutions

were diluted and measured by ICP-MS.

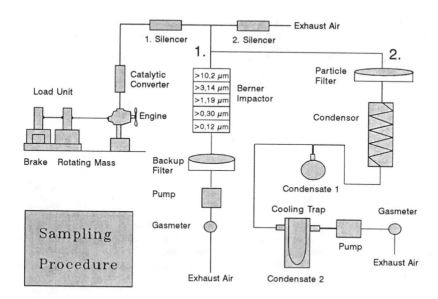

<u>Figure 1</u>   Sampling procedure

2) The glassfibre filters were digested following the modified method MDHS 46 (Methods for the Determination of Harzardous Substances, July 1985, Platinum metal and soluble inorganic compounds of platinum in air). This moderate treatment by MDHS 46 dissolves only a fraction of the total platinum, which is usually called "soluble" platinum. Therefore the filters were cut into smaller pieces and dissolved in 0,1 M HCl. These solutions were placed in an ultrasonic bath for 10 minutes. The soluble content was filtered by a fine filter (porous size: 2,2 μm) and prepared by adding de-ionised water. The bulk of the collected platinum, however, remains on the glassfibre filters and is dissolved in a subsequent step with aqua regia. The solutions were evaporated almost to dryness and the residue re-dissolved twice in 2 M HCl and at last in 0,1 M HCl. The final sample solution which contained the "insoluble" platinum was prepared by adding de-ionised water.

From the condensate samples water was evaporated by a rotary evaporator. These solutions were transferred quantitatively to the high pressure asher and were digested as described above.

All samples were determined by ICP-MS.

<u>Reagents</u>

All reagents were of high-purity grade. The acids were prepurified by surface-distillation [3].

<u>Operating Parameters of ICP-MS</u>

Instrument: PlasmaQuad 2+, Fisons Instruments
Nebulization: Gilson peristaltic pump, uptake rate 1 ml/min; glass concentric nebulizer type TR-30-A3; scott double pass spray chamber, water cooled to 10°C.
Plasma: Fassel type torch; RF frequency 27 MHz; Plasma forward RF power 1,35 kW; Plasma reflected RF power 5 kW.
Gas flows: Coolant Ar flow: 13 l/min; auxiliary Ar flow 0,6 l/min; nebulizer Ar flow 0,889 l/min; sample uptake rate 1,0 l/min.
Cones: Sampling cone made of Ni 1,0 mm orifice; skimmer cone made of Ni 0,7 mm orifice; ion lens setting optimised on Lu (mass 175).
Peak Jump conditions: Mass range 100 to 200 amu; number of channels 2048; number of scan sweeps 100; dwell time 320 μs; points per peak 5; dac step between point 5; number of peak jump sweeps 20; collector type pulse; Pt isotopes 194, 195, 196.

3 DISCUSSION AND RESULTS

In our study twelve three-way catalytic converters are tested under varied conditions. The cake represents a typical particle distribution of one Berner Impactor experiment (figure 2). About 70 percent of the particles have a diameter larger than 10.2 μm. The particles with diameters between 10.2 and 0.12 μm are demonstrated in the column. Only 6 percent of all particles are smaller than 0.30 μm.

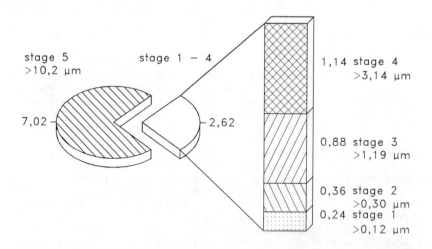

<u>Figure 2</u>   Typical particle distribution of platinum in ng/m³;
                      simulated speed 80km/h

In figure 3 the results of the catalytic converters of production line B are outlined. For every particle size three columns are shown. Each column represents the particulate platinum emission in ng/m³. Every concentration indicated is the average of six individual experiments of one catalytic converter: The column on the left is named B1, the column in the middle B2 and the column on the right B3. On one hand three converters of the same production line and on the other hand the results of two different driving conditions are compared. It is obvious that the emissions of B1, B2 and B3 are different. The most striking effect is shown on stage 4 in figure 3b. B2 and B3 emit about 1 ng/m³ platinum particles while B1 emits five times more platinum. A difference of a factor of 2 between the results of the converters is in fact usual, looking at stage 5 in figure 3b or at stage 3-5 in figure 3a. Comparing the emissions under different driving conditions the platinum concentration of the constant speed experiment are about 4 times lower. The concentrations which are emitted range from 0.4 (stage 1) - 20 (stage 5) ng/m³ under cycle conditions and from 0.2 (stage 1) - 5 (stage 5) ng/m³ with constant speed. This effect can be explained by the extreme changes of the exhaust flow rate and temperatures which are the consequence of more mechanical abrasion under cycle conditions.

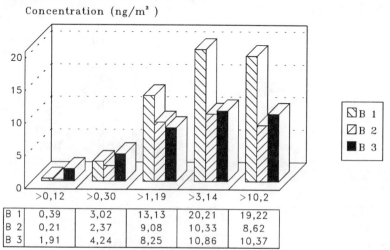

B 1	0,39	3,02	13,13	20,21	19,22
B 2	0,21	2,37	9,08	10,33	8,62
B 3	1,91	4,24	8,25	10,86	10,37

Particle Size (μm)

Figure 3a   Particle distribution of platinum (ng/m³) from the catalytic converters B1, B2, B3; US 75 Driving Cycle

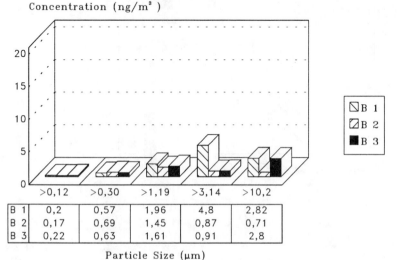

B 1	0,2	0,57	1,96	4,8	2,82
B 2	0,17	0,69	1,45	0,87	0,71
B 3	0,22	0,63	1,61	0,91	2,8

Particle Size (μm)

Figure 3b   Particle distribution of platinum (ng/m³) from the catalytic converters B1, B2, B3; simulated speed 80 km/h

In figure 4a and 4b the platinum emissions of different manufacturers are demonstrated. The particle size distribution of the converters of production line C are completely different from the results of the converters of production line B. Nearly all particles of line C are larger than 10.2 μm while the particles of line B are well distributed over the five stages of the impactor. The particulate emission, however, demonstrate that these two converters are produced by different procedures.

In figure 5 the total particle emission of platinum for the four manufacturers is shown. Figure 5a represents the cycle results, figure 5b the results of constant speed. Again the emissions under cycle conditions (nearly 60 ng/m^3) are higher than under constant speed (about 10 ng/m^3). The difference between the three converters of the same production line are under cycle conditions larger than with constant speed. The reason

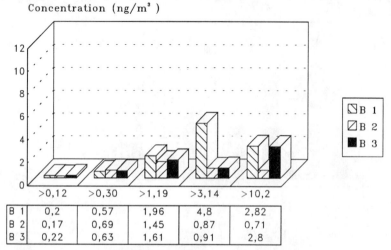

Concentration (ng/m^3)

B	>0,12	>0,30	>1,19	>3,14	>10,2
B 1	0,2	0,57	1,96	4,8	2,82
B 2	0,17	0,69	1,45	0,87	0,71
B 3	0,22	0,63	1,61	0,91	2,8

Particle Size (μm)

Figure 4a   Particle distribution of platinum (ng/m^3) from the converters of production line B; simulated speed 80 km/h

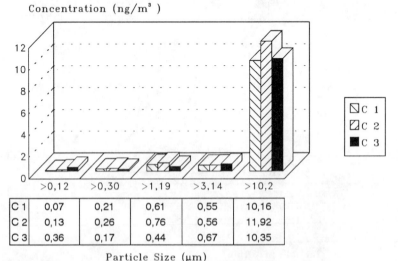

Concentration (ng/m^3)

C	>0,12	>0,30	>1,19	>3,14	>10,2
C 1	0,07	0,21	0,61	0,55	10,16
C 2	0,13	0,26	0,76	0,56	11,92
C 3	0,36	0,17	0,44	0,67	10,35

Particle Size (μm)

Figure 4b   Particle distribution of platinum (ng/m^3) from the converters of production line C; simulated speed 80 km/h

for this effect is the possibility to make isokinetic experiments under constant speed while under cycle conditions this is not possible. A critical orifice regulates the volumetric flowrate in the exhaust line, under cycle conditions it is only possible to take an orifice for the average speed. Remarkable is the fact that the total platinum emission of the different converters A,B,C and D is more or less (exception, figure 5a converter B1) the same although the production process and the particle distribution over the impactor of each converter is different.

Concentration (ng/m³)

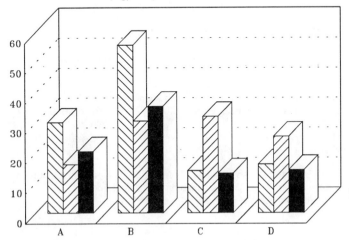

Figure 5a   Total particle emission of platinum of four different manufacturers A,B,C,D; US 75 Driving Cycle

Concentration (ng/m³)

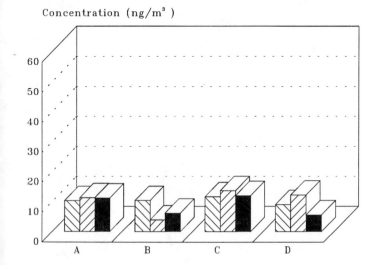

Figure 5b   Total particle emission of platinum of four different manufacturers A,B,C,D; simulated speed 80 km/h

4 CONCLUSION

Continuing the experiments, recently published [4], we try to put the results on a broadened, statistically ensured basis. The presented results show that every converter is an individual one, so that the platinum emissions can differ up to a factor of 5. But the total platinum emissions have more or less the same order of magnitude. The results given here are derived from new catalytic converters, in the future we will test middle-old and old converters, too.

REFERENCES

1.   Kathleen C. Taylor, "Automobile Catalytic Converters", Springer Verlag Berlin Heidelberg New York Tokyo 1984.

2.   J.W. Hamersma, S.L. Reynolds and R.F. Maddalone, "IERL-RTP Procedures Manual: Level I Environmental Assessment. US EPA" 600/2-76-106a, 1977.

3.   E.C. Kuehner, R. Alvarez, P.J. Paulsen and T.J. Murphy, Anal. Chem. **44**, 2050 (1972).

4.   H.P. König, R.F. Hertel, W. Koch and G. Rosner, "Determination of Platinum Emissions from a Three-way Catalyst-Equipped Gasoline Engine", Atmospheric Environment Vol. 26A, **5**, 741-45 (1992).

Determination of Trace Elements in 4 Chinese Bio-geochemical Reference Materials by ICP-MS

Yin Ming and Yin Ningwan
INSTITUTE OF ROCK AND MINERAL ANALYSIS, 26 BAIWANZHUANG
DAJIE, BEIJING 100037, CHINA

More than 40 trace elements in 4 Chinese bio-geochemical
reference materials GBW 07602 and GBW 07603 (bush leaves and
branches), GBW-07604 (poplar leaves) and GBW 07605 (tea
leaves) were determined by Inductively Coupled Plasma Mass
Spectrometry (ICP-MS). A microwave sample dissolution technique with-out using $HClO_4$ and/or H_2SO_4, which are known to
produce various molecular species that overlap several
important isotopes in ICP-MS analysis, was used. The results
obtained in this work are generally in good agreement with
the certified values from sub-ppm to hundreds of ppm, and
illustrate the versatility of the technique, particularly for
some less well-characterized elements.

Introduction

A set of 4 Chinese bio-geochemical standard reference
materials GBW 07602-07605, was prepared and issued by the
Institute of Geophysical and Geochemical exploration (IGGE).
The original purpose was to use these materials as primary
standards for the quality monitoring of geochemical vegetation survey in some prospecting programs. But their use has
been extended to other fields, such as agriculture and eco-
environmental sciences. 21 laboratories in China have joined
in a collaborative study to analyse these materials using a
variety of analytical methods and instrumentations including
ICP-AES, INAA, XRF, AAS, ICP-AFS, colorimetry and polaro-
graphy.
Inductively Coupled Plasma Mass Spectrometry (ICP-MS) is a
newly developed analytical method with low detection limits
for most elements, high dynamic range, high throughput and
the capability of simultaneous multi-element analysis[1-3].
ICP-MS is, thus, considered to be particularly suitable for
such work.

Some previous studies on the ICP-MS analysis of biological
and plant materials have utilized time-consuming wet
-digestion procedure[4,5] which leads to a lengthy and
tedious operation and takes the risk of contamination.In most
cases, H_2SO_4 and/or $HClO_4$ are involved to get complete

decomposition of the organic materials. It doesn't seem to
advisable in ICP-MS analysis, because these acids are know
to produce various molecular species which overlap wi
several important isotopes[6]. Recent introduction of the micrc
wave dissolution technique without using H_2SO_4 and HClO
appears to be a promising method for the pretreatment of t
organic materials[7].

In this paper, the concentration of more than 40 tra
elements in 4 Chinese bio-geochemical reference materials G
07602- 07605 were determined by ICP-MS after microwave di
solution. The results obtained in this work are generally
good agreement with the certified values.

Experimental

Apparatus:
The instrument employed in this work was a PlasmaQu
Inductively Coupled Plasma Mass-Spectrometer (VG Elementa
Winsford, UK).The operating conditions are listed in Table

A conventional microwave oven (FEIYUE Model WL 5001, 500
Shanghai, China) containing a rotating turntable with s
sample positions, was used for the sample digestion.

Reagents and standard solutions for instrument calibratio
Unless stated otherwise, all reagents were of analytica
reagent grade. Doubly distilled de-ionized water was us
throughout.

Multi-element standard solutions were prepared in 2% (v/
nitric acid by dilution and mixing of 1g/L stock solutions
individual elements. Elements mixed were divided into t
groups in order to avoid the interaction between sc
elements and reagents. The elements chosen for measuremer
in each standard solution are listed in Table 2. Standa
solutions were spiked with 100µg/L of Rh as an interr
standard.

Sample preparation
Samples were air-dried to constant weight in a drying o
at 80 °C for four hours. Dried samples of ca.0.1 g we
weighed accurately into the Teflon high-pressure microwa
acid-digestion bomb (Fei Yue, Analytical Instrument Facto
Shanghai, China). 1 ml of 70% HNO_3 and 1 ml of 30% H_2
were added. The bombs were sealed tightly and then positio
in the microwave carousel. The system was operated at f
power for six minutes. After cooling, the digest was quan
tatively transferred to a 10 ml test-tube. The solution
spiked with 100 ng/ml Rh as an internal standard and fina

Table 1. Operating Conditions

Power	1.25 kw	Expansion stage	2.7 mbar
Nebulizer gas flow	0.750 1 min^{-1}	Intermediate stage	$1.0*10^{-4}$ mbar
Auxiliary gas flow	1.1 1 min^{-1}	Analyser stage	$3.0*10^{-6}$ mbar
Plasma gas flow	14 1 min^{-1}		
Sample uptake rate	1.2 ml min^{-1}	Scan conditions:	
Sampling depth		Mass range	6–240 amu
above Load coil	10 mm	Numbers of scan sweeps	120
Spray chamber		Dwell time	500 us
cooled to	10 C	Number of channels	2048

Table 2. The elements in two calibration standard solutions

SS–A: Li,Be,B,Ti,Cr,Mn,Co,Ni,Cu,Zn,Ga,Rb,Sr,Y,Zr,Nb,Mo,Cd
        Sb,Cs,Ba,La,Ce,Pr,Nd,Sm,Eu,Gd,Tb,Dy,Ho,Er,Tm,Yb,Lu,
        Hg,Tl,Pb,Bi,Th,U.
SS–B: V,As,Ag.

made up to 10 ml. Each sample was prepared with five
replicates.

ICP-MS measurements

Since the elements of interest cover the whole mass range, the
ICP-MS was used in the full spectrum scanning mode. The scan
conditions are summarized in Table 1. Each measurement lasted
for ca.123 s. In order to avoid memory effects, the sample
introduction system was rinsed with 2% (v/v) HNO_3 for 2 mins
between different sample solutions.

Results and Discussion

Results ($\mu g/g$) for the elements of interest (mean standard
deviation) are presented in Table 3, with certified and
uncertified values where available, and their confidence
limits.

The isotopes used for the analysis (see Table 4) were
carefully selected to avoid possible isobaric and polyatomic
ion interferences. However, for As and V in GBW 07602 and GBW
07603, the measured values were a little bit higher than
certified values. These positive errors could come from polyatomic ions of ArCl and ClO (Concentration of Cl in these two
samples are 1.13% and 1.92% respectively). Somewhat similar
results were also observed on Eu in GBW 07604 and GBW 07605
due to the interferences of BaO on Eu (concentration ratios
of Ba/Eu in these two samples are ca.2800 and 3200
respectively). No correction was made for Eu because it is
difficult to make a reliable correction for it at such low
concentration level.

Detection limits (DL) were defined as the concentration of
the analyte which would produce a signal equivalent to three
times the standard deviation of the background obtained
for a 2% (v/v) HNO_3 blank solution. Quantitation limits (QL)
were calculated as 10 times the SD of these measurements. The
DL and QL for all elements determined are given in Table 4.
It is notable that the DL and QL are significantly low for
most element determined in this work. But the analysis of Se
is limited by a substantial interference from Ar with
principal and secondary isotopes at m/z=80 and 78, respectively. The remaining isotopes have natural abundances below 10%.
It is impossible to obtain reliable results for Se by using
isotopes of such low abundances.

The concentrations of Be in most samples were lower than its
detection limit of the method, but for those elements with
concentrations around their quantitation limits, there was a

Table 3. Concentrations (µg/g) of trace elements in 4 Chinese bio-geochemical Standard Reference Materials (mean ± standard deviation)

Sample	GBW 07602 (灌木)		GBW 07803 (灌木)		GBW 07804 (杨树叶)		GBW 07805 (茶叶)	
Element	Found	Certified	Found	Certified	Found	Certified	Found	Certified
Li	2.1±0.4	2.4±0.4	2.3±0.2	2.6±0.4	1.1±0.1	0.84±0.15	----	(0.38)
Be	0.041±0.007	0.056±0.014	0.052±0.010	0.051±0.004	0.033±0.007	0.021±0.005	0.023±0.007	0.034±0.006
B	31±3	34±7	33±1	38±6	51±4	53±5	12±1	15±4
Ti	32±4	95±18	30±2	95±20	12±0.8	20.4±2.2	12.2±2	24±4
V	3.8±0.4	2.4±0.3	3.6±0.1	2.4±0.4	0.94±0.005	(0.64)	0.65±0.01	(0.86)
Cr	2.8±0.4	2.3±0.3	2.5±0.3	2.8±0.2	1.3±0.2	0.55±0.07	1.8±0.2	0.80±0.03
Mn	60±5	58±6	60±2	61±5	43±3	45±4	676±8	1240±70
Co	0.38±0.02	0.39±0.05	0.38±0.02	0.41±0.05	0.39±0.02	0.42±0.03	0.16±0.01	0.18±0.02
Ni	1.8±0.2	1.7±0.4	1.4±0.1	1.7±0.3	1.7±0.1	1.9±0.3	4.4±0.4	4.6±0.5
Cu	5.0±0.6	5.2±0.5	6.1±0.4	6.6±0.8	8.4±1.0	9.3±1.0	16±2	17.3±1.8
Zn	19.1±1.6	20.6±2.2	47±2	55±4	32±2	37±3	24±2	26.3±2.0
Ga	0.46±0.04	----	0.36±0.02	----	0.35±0.05	----	0.62±0.06	----
As	1.42±0.2	0.95±0.12	2.0±0.1	1.25±0.15	0.45±0.06	0.37±0.09	0.28±0.07	0.28±0.04
Rb	3.5±0.3	4.2±0.2	3.7±0.2	4.5±0.6	8.4±0.4	7.6±0.8	80±4	74±5
Sr	330±20	345±11	231±9	248±18	150±7	154±9	14.2±0.4	15.2±0.7
Y	0.66±0.06	(0.83)	0.80±0.03	0.88±0.02	0.13±0.01	0.145±0.015	0.30±0.02	0.38±0.04
Zr	0.28±0.03	----	0.29±0.02	----	0.15±0.01	----	0.14±0.04	----
Nb	0.079±0.006	----	0.075±0.01	----	0.023±0.002	----	0.016±0.002	----
Mo	0.24±0.01	0.28±0.04	0.24±0.03	0.28±0.05	0.18±0.02	0.18±0.01	0.041±0.01	0.038±0.007
Ag	0.035±0.007	0.027±0.006	0.059±0.008	0.049±0.007	0.012±0.003	(0.013)	0.017±0.003	(0.018)
Cd	0.20±0.02	0.14±0.08	0.65±0.07	(0.38)	0.36±0.002	0.32±0.07	0.052±0.010	0.057±0.010
Sb	0.063±0.010	0.078±0.020	0.080±0.010	0.095±0.014	0.045±0.01	0.045±0.006	0.046±0.010	0.058±0.006

Table 3. Cont.

Cs	0.25±0.01	0.27±0.03	0.24±0.01	0.27±0.02	0.050±0.005	0.053±0.003	0.28±0.01	0.29±0.02
Ba	16±1	19±3	14±1	18±2	24±1	26±4	59±2	58±6
La	1.3±0.2	1.23±0.10	1.1±0.1	1.25±0.06	0.23±0.01	0.26±0.02	0.55±0.03	0.60±0.04
Ce	2.1±0.1	2.4±0.3	2.0±0.1	2.2±0.1	0.45±0.03	0.49±0.07	0.88±0.04	1.0±0.2
Pr	0.27±0.02	-----	0.26±0.02	(0.24)	0.055±0.009	-----	0.12±0.01	(0.12)
Nd	1.0±0.05	(1.1)	1.0±0.1	1.0±0.1	0.17±0.02	(0.22)	0.38±0.02	(0.44)
Sm	0.17±0.02	0.19±0.01	0.18±0.01	0.19±0.02	0.046±0.005	0.038±0.006	0.062±0.015	0.085±0.023
Eu	0.040±0.008	0.037±0.002	0.040±0.006	0.039±0.003	0.017±0.004	0.009±0.003	0.029±0.003	0.018±0.002
Gd	0.16±0.01	-----	0.17±0.02	(0.19)	0.041±0.007	(0.043)	0.075±0.015	(0.093)
Tb	0.022±0.002	(0.028)	0.021±0.002	0.025±0.003	0.004±0.001	-----	0.010±0.002	(0.011)
Dy	0.15±0.01	-----	0.18±0.02	(0.13)	0.037±0.01	(0.036)	0.043±0.010	(0.074)
Ho	0.033±0.004	-----	0.030±0.002	(0.033)	0.007±0.001	-----	0.015±0.004	(0.019)
Er	0.041±0.002	-----	0.045±0.005	-----	0.012±0.002	-----	0.024±0.008	-----
Tm	0.006±0.001	-----	0.007±0.001	-----	0.002±0.0008	-----	0.005±0.001	-----
Yb	0.053±0.006	0.063±0.014	0.056±0.007	0.063±0.009	0.010±0.002	0.018±0.004	0.040±0.006	0.044±0.005
Lu	0.008±0.002	-----	0.010±0.001	(0.011)	0.004±0.0005	-----	0.005±0.0015	(0.007)
Hg	-----	-----	-----	-----	0.080±0.010	0.026±0.003	0.050±0.008	(0.013)
Tl	0.015±0.002	-----	0.032±0.003	-----	0.010±0.002	-----	0.021±0.001	-----
Pb	6.9±0.4	7.1±1.1	46.2±3	47±3	1.3±0.08	1.5±0.3	3.6±0.2	4.4±0.3
Bi	0.021±0.002	(0.022)	0.021±0.003	0.023±0.005	0.020±0.002	0.027±0.002	0.080±0.003	0.083±0.008
Th	0.34±0.02	0.37±0.02	0.35±0.02	0.38±0.04	0.070±0.005	0.070±0.01	0.045±0.006	0.081±0.009
U	0.093±0.003	(0.11)	0.11±0.005	(0.12)	0.023±0.003	(0.028)	0.014±0.001	-----

Mean values were calculated from single determinations of five separate sample preparations.

( ) uncertified values.

----- not detected.

Certified values are from IGGE Yan Mingcai et al. (1991)

Table 4. Detection limits (DL) and Quantitation limts (QL) in 2% HNO_3 solution (ug/ml)

Element	Mass	DL	QL	Element	Mass	DL	QL
Li	7	0.2	0.7	Cs	133	0.045	0.15
Be	9	0.2	0.7	Ba	138	0.02	0.07
B	11	0.8	2.7	La	139	0.03	0.1
Ti	47	0.5	1.7	Ce	140	0.04	0.13
V	51	0.23	0.77	Pr	141	0.03	0.1
Cr	52	0.05	0.17	Nd	146	0.15	0.5
Mn	55	0.07	0.23	Sm	147	0.1	0.33
Co	59	0.05	0.17	Eu	153	0.04	0.13
Ni	60	0.2	0.7	Gd	157	0.08	0.27
Cu	63	0.12	0.4	Tb	159	0.03	0.1
Zn	66	0.1	0.33	Dy	163	0.07	0.23
Ga	69	0.1	0.33	Ho	165	0.02	0.07
As	75	0.2	0.7	Er	166	0.05	0.17
Rb	85	0.1	0.33	Tm	169	0.02	0.07
Sr	88	0.08	0.27	Tb	174	0.1	0.33
Y	89	0.03	0.1	Lu	175	0.01	0.033
Zr	90	0.1	0.33	Hg	202	0.08	0.27
Nb	93	0.06	0.2	Tl	203	0.03	0.1
Mo	98	0.14	0.47	Pb	208	0.16	0.54
Ag	107	0.07	0.23	Bi	209	0.03	0.1
Cd	114	0.3	1.0	Th	232	0.02	0.07
Sb	123	0.13	0.43	U	238	0.02	0.07

DL=3*SD; QL=10*SD; (SD=standard deviation of blank signals)

good agreement between experimental and certified values. even below quantitation limits, the data of Ag were good enough to be accepted.

There is a series of phenomena, termed matrix effects, which have been observed in the ICP-MS analysis[8]. These phenomena appear to be dependent on the total dissolved salt in the sample solution being nebulized. The most severe one can result in clogging of nebulizer and/or of sampler and skimmer orifices due to high total dissolved solids in the sample solution. Although the total dissolved solids in solutions in this work were below 0.1% signal drift with time was still observed during the course of analysis, typically ca. 20% signal depression can be observed within one hour continuous aspiration of sample solution. Table 5 summarizes the results for some elements from ICP-MS analysis of GBW 07602 with different sample dilution factors. The elements chosen span the whole mass range.The excellent agreement from the sample solutions with different dilution factors indicates that the matrix effects can be adequately compensated by using Rh as internal standard.

The discrepancy was very noticeable on Mn in GBW 07605. This is simply because the concentration and hence the signal of Mn was so high that the detector became saturated.The further dilution of the sample can improve the result, however, this approach will not be regarded as appropriate in cases where simultaneous multielement analysis is a major concern.

The values for Ti in this work are significantly lower than the certified values. The possible cause is considered to be incomplete decomposition of Ti mineral in the samples by proposed method. This has been proved by using HNO_3+HClO_4+HF mixed acid digestion procedure in which the good recovery of Ti was obtained. The values Hg in GBW 07604 are higher than certified values, possibly due to contamination.

Most of the data obtained by ICP-MS in this work were in excellent agreement with the certified value and the precision for most elements was better than 10% RSD.

Conclusion

This work has shown that accurate determination of trace elements in bio-geochemical samples (vegetations), even at sub $\mu g \ g^{-1}$ levels can be accomplished by ICP-MS. A rapid and complete dissolution of the vegetation materials was achieved using a microwave digestion procedure requiring only concentrated nitric acid and hydrogen peroxide. The elimination of

Table 5. Results (µg/ml) obtained by ICP-MS for 7 elements in GBW 07602 with different dilution factors (DF), using Rh as an internal standard (mean standard deviation of 3 measurements)

Element	DF = 100		DF = 300		DF = 1000	
[11] B	2.1	0.4	2.1	0.3	2.2	0.4
[55] Mn	60	5	59	3	61	3
[66] Zn	19.1	1.6	20	2	19.6	1.7
[88] Sr	330	20	342	18	333	14
[138] Ba	16	1	16	1	17	1.4
[140] Ce	2.1	0.1	2.2	0.2	2.2	0.3
[208] Pb	6.9	0.4	6.7	0.4	7.2	0.6

hydrochloric acid and perchloric acid from the digestion allows the elements, such as As and V, which suffer from polyatomic ions of chlorine,to be determined.

References

1. Ridout,P.S., Jones,H.R., and Williams,J.G., Analyst, 1988 113, 1383.
2. Gray,A.L., Spectrochim Acta, part B, 1985, 40,1525.
3. Houk,R.S., Anal. Chem., 1986, 58, 79A.
4. G.F.Clements., J.Radioanal.Chem. 1976, 32, 25.
5. H.R.Roberts., Food Safety (Wiley, New York, 1981),
6. M.A.Vaughan and G.Horlick, Appl. Spectrosc. 1986 40, 434.
7. James K.Friel,Craig S.Skinner, Simon E.Jackson and Henry P.Longerich, Analyst 1990, 115, 269.
8. W.Doherty and A.Vander Voet. Can.J.Spectra. 30 (6) 1985.

A Comparative Study of ICP-MS and TIMS for Measuring Skin Absorption of Lead

L. S. Dale, J. L. Stauber, O. P. Farrell, and T. M. Florence
CSIRO DIVISION OF COAL AND ENERGY TECHNOLOGY, LUCAS
HEIGHTS, N.S.W., 2234, AUSTRALIA

B. L. Gulson
CSIRO DIVISION OF EXPLORATION GEOSCIENCE, NORTH RYDE,
N.S.W., 2113, AUSTRALIA

1 INTRODUCTION

The major sources of lead exposure are ingestion and inhalation. Exposure by absorption through the skin was previously thought to occur only if the lead was present as lipid-soluble organic complexes such as tetraethyl lead. It has however, been shown that many inorganic forms of lead such as lead metal, lead oxide and lead nitrate are rapidly absorbed through the skin[1]. An absorption mechanism was proposed based on the rapid diffusion of lead ions through the filled sweat ducts and their slower diffusion through the stratum corneum into the blood capillaries[2]. It was found that approximately 1.5 percent of the total lead absorbed through the skin was excreted in sweat. Similar amounts were found in sweat when lead was introduced intravenously[3] or by inhalation or ingestion[4].

The behaviour of skin-absorbed lead is very different from ingested lead. Previous skin absorption studies using the enriched stable isotope [204]Pb in which TIMS was used to analyse blood samples, revealed that, while there was a substantial increase in [204]Pb in blood, no significant increase in the total lead in blood was observed. Since blood lead is the main criterion by which industry determines lead exposure, skin absorption of lead would remain undetected. The possibility remains that a significant proportion of the lead body burden of workers results from skin absorption. The occupational hygiene implications of this are profound. Although precautions are taken in the lead industry to protect workers from ingestion and inhalation of lead, no attempt is made to avoid skin contact.

To further investigate the fate and transport of skin-absorbed lead, a controlled experiment was carried out using lead enriched in [204]Pb. Sweat and urine samples were analysed by both ICPMS and TIMS. The rationale behind the use of both techniques was that ICPMS, because of its high sample throughput and minimal sample preparation requirements, would provide rapid preliminary data on the lead transport. At the time it was not known whether these data would be adequate to provide a satisfactory monitoring profile. On the other hand, analysis of the samples by TIMS, because of its established accuracy and precision, was expected to provide a more definitive indication of the lead transport.

The experiment, therefore, provided the opportunity to compare the analytical performance of ICPMS and TIMS in relation to isotope ratio data and determination of low levels of lead in body fluids.

2   EXPERIMENTAL

Preparation of ^{204}Pb (CH$_3$COO)$_2$

Lead metal (4.8 mg) obtained from Harwell and containing 49 at. percent ^{204}Pb, was dissolved in 0.3 mL concentrated nitric acid (Merck Suprapur) by repeated evaporation to dryness. Milli-Q water (60 µL) and 15 µL 1M acetate buffer (pH 4.7) were added to bring the final solution to pH 4.

Lead Exposure

The left arm of a subject was washed with 5 percent Extran-300 (BDH) detergent, rinsed with Milli-Q water and dried with a Whatman 542 filter paper.

The Pb (CH$_3$COO)$_2$ solution was pipetted onto a 25mm Millipore HATF (0.45 µm) membrane filter and placed on the flexor surface of the subject's left arm. It was covered with a small piece of acid-washed Parafilm and held in place for 24 hours on the forearm using clear household polyethylene film.

Collection of Sweat Samples

Sweat samples from the opposite arm were collected in a sauna the day before exposure and $1^1/_2$, $3^1/_4$, 6 and 25 hours after the application of the lead. At the onset of sweating the subject washed the right arm with Milli-Q water. Sweat was then collected from this arm into acid-washed Savillex containers, weighed and stored at 4°C prior to analysis.

Collection of Urine Samples

Urine samples were collected in 500 mL acid-washed polyethylene containers before lead exposure and $1^1/_2$, $3^1/_4$, 6, 15, 25 and 47 hours after lead exposure. The samples were split for analysis by ICPMS and TIMS.

Analysis by ICPMS

Sweat samples were diluted five-fold with Milli-Q water. Urine samples were diluted ten-fold. They were acidified with Merck Suprapur nitric acid (1 percent $^w/_w$) and spiked with bismuth internal standard solution at a concentration equivalent to 100 µg L^{-1}. Calibration standards of lead in the range 0.1 to 5 µg L^{-1} were prepared containing the same acid strength and bismuth concentration as the samples.

The samples and standards were run on a VG
Plasmaquad PQ2 PLUS (Fisons Instruments) using an
isotope ratio peak jump procedure. The conditions are
given in Table 1.

Table 1 ICPMS Instrumental Conditions

Measurement Mode	peak jump
Dwell Time	10240 µs
Points per Peak	7
DAC Steps	5
No. of Sweeps	120
No. of Runs	10
Isotopes Measured	204, 206, 207, 208, 209
Sensitivity	10 M cps / ppm equivalent

Analysis of TIMS

Sweat and urine samples were microwave digested
with sub-boiling distilled nitric acid. Each sample was
spiked with a standard ^{202}Pb solution prior to digestion
to enable measurement of lead content and isotope ratios
on the same solution. The lead in the samples was
separated using a two-stage ion exchange procedure. The
initial separation used a 0.5 cc bed of AG1-X4 200-400
mesh anion exchange resin and final separation
purification used a 0.05 cc bed of AG1-X8 200-400 mesh
anion exchange resin. All operations were carried out
in laminar flow work bench stations.

Purified samples were loaded onto outgassed rhenium
filaments using silica gel/phosphoric acid as an
emitter. The samples were run on a VG Sector 54 Mass
Spectrometer (Fisons Instruments) using the conditions
given in Table 2.

Table 2 TIMS Instrumental Conditions

No. of Scans	60
Integration Time	1 s per peak, 2 s delay
Run Time	10 min after beam stabilisation
Ratios Measured	$^{208}/_{206}$, $^{207}/_{206}$, $^{206}/_{204}$

3   RESULTS AND DISCUSSION

The results for the concentrations of ^{204}Pb and total
Pb in the sweat and urine samples obtained by ICPMS and
TIMS are shown in Table 3. Agreement for the sweat
samples is considered very satisfactory given that the
actual levels measured by ICPMS are one-fifth lower than
indicated. The agreement obtained for the urine samples
is, in some cases, not as good. The reason for this was
later attributed to the presence of mercury which was
detected in the residue of one sample after the isotope
ratio measurements had been performed. Mercury would
have a significant contribution to mass 204 thus
affecting the ^{204}Pb determination. This was most

<u>Table 3</u> Concentrations of ^{204}Pb and Total Pb in Sweat (S) and Urine (U) Samples

SAMPLE	TIME* (h)	ICPMS ^{204}Pb µg L^{-1}	ICPMS Total Pb µg L^{-1}	TIMS ^{204}Pb µg L^{-1}	TIMS Total Pb µg L^{-1}
S1	0	0.46	31	0.63	44
S2	$1^1/_2$	0.17	9.9	0.17	9.5
S3	3	0.61	5.5	0.70	5.4
S4	6	0.51	9.3	0.60	8.3
S5	25	6.7	24	9.1	27
U1	0	0.20	9.3	0.09	6.4
U2	$1^1/_2$	0.17	4.3	0.16	4.8
U3	$3^1/_4$	0.13	6.7	0.08	4.7
U4	6	0.16	8.6	0.12	7.2
U5	15	0.35	18	0.07	4.7
U6	25	0.13	6.0	0.09	5.7
U7	47	0.13	5.0	0.05	3.1

* Time sample taken after application of lead.

evident in the ICPMS results for ^{204}Pb in samples U1 and U5.

Of greater significance is the lead concentration profiles inferred from the ^{204}Pb results. For the sweat, there was only a significant increase in ^{204}Pb after 25 h exposure. For urine, no significant breakthrough of ^{204}Pb was apparent. No significant trends in the sweat and urine samples were observed even when the concentrations of both total Pb and ^{204}Pb concentrations were normalised to chloride and creatinine respectively.

In contrast to this the results obtained for the ^{204}Pb isotope abundances (expressed as atomic percent ^{204}Pb) in the samples shown in Table 4, indicate a sharp breakthrough after 3 hours followed by gradual increase to 25 hours for sweat and a sharp breakthrough after $1^1/_2$ hours for urine. The fact that these profiles were not detectable from ^{204}Pb and total Pb concentrations is indicative of the complex mechanisms of lead transport in the body and highlights the uncertainty in monitoring lead body burden.

A comparison of the ICPMS and TIMS data for ^{204}Pb isotope abundances in sweat and urine are shown in Figures 1 and 2 respectively. The profiles obtained by ICPMS for both the sweat and urine profiles were very similar to those obtained with the TIMS data. The ICPMS data for sweat appeared to exhibit some bias. This is considered of minor consequence since the same profile was obtained.

Of perhaps greater significance is the relative ease with which the ICPMS data was obtained. Using minimal sample preparation and a simple calibration procedure the data was obtained and processed in one day. This is in contrast to the need for lengthy sample pretreatment

(3-4 days) and instrument setup time for TIMS.   Thus the
ICPMS technique is extremely cost-effective.

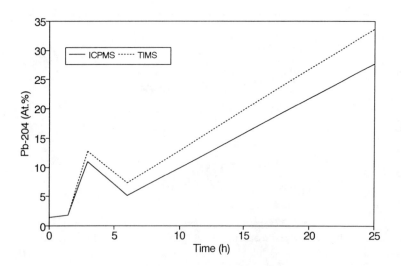

Figure 1   Comparison between ICPMS and TIMS results for
[204]Pb (at. percent) in sweat.

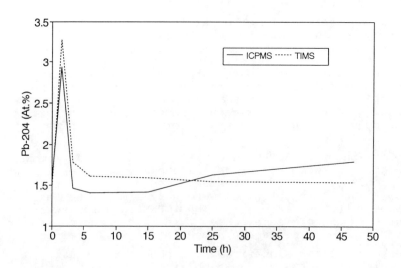

Figure 2   Comparison between ICPMS and TIMS results for
[204]Pb (at. percent) in urine.

TABLE 4 ^{204}Pb Abundances in Sweat and Urine Samples

^{204}Pb (at. percent)

SAMPLE	TIME (h)	ICPMS ($\pm 2\sigma$)	TIMS ($\pm 2\sigma$)
S1	0	1.481±0.027	1.4473±0.0022
S2	$1^1/_2$	1.754±0.054	1.7674±0.0027
S3	3	11.06±0.31	12.850±0.019
S4	6	5.15±0.39	7.249±0.011
S5	25	27.66±0.066	33.738±0.051
U1	0	1.55±0.16	1.457±0.002
U2	$1^1/_2$	2.94±0.20	3.279±0.005
U3	$3^1/_4$	1.46±0.08	1.775±0.003
U4	6	1.40±0.14	1.606±0.002
U5	15	1.41±0.07	1.587±0.002
U6	25	1.63±0.13	1.541±0.002
U7	47	1.79±0.07	1.527±0.002

4   CONCLUSIONS

The use of lead enriched in ^{204}Pb coupled with isotope ratio measurements has provided important data on the transport of lead absorbed in the skin. It has been shown that ICPMS has the capability of providing data adequate for monitoring lead transport at the very low levels of concentration encountered in body fluids. Although ICPMS lacks the accuracy and precision achieved by TIMS, its application described here is enhanced by the minimal sample preparation required and the high sample throughput than can be achieved.

REFERENCES

1.   T.M. Florence, S.G. Lilley and J.L. Stauber, The Lancet, 1988, July, 157.

2.   S.G. Lilley, T.M. Florence and J.L. Stauber, Sci. Total. Environ., 1988, 76, 267.

3.   B.C. Campbell, P.A. Meredith, M.R. Moore and W.S. Watson, Toxicol. Lett., 1984, 21, 231.

4.   M.B. Rabinowitz, G.W. Wetherill and J.D. Kopple, J. Clin. Invest., 1976, 58, 260.

ACKNOWLEDGEMENT

We wish to thank Karen Mizon for some of the TIMS measurements and Greg Kilby who volunteered for the lead uptake experiment.

Fundamental Aspects of an Analytical Glow Discharge

M. van Straaten and R. Gijbels
UNIVERSITY OF ANTWERP, DEPARTMENT OF CHEMISTRY,
UNIVERSITEITSPLEIN I, B-2610 WILRIJK-ANTWERP, BELGIUM

I. Introduction.

In the field of inorganic solid analysis, mass spectrometric methods are shown to combine important advantages: they are the only universal multi-element methods which allow the determination of all elements with the required detection limits. Also the simplicity of the obtained spectra (compared to optical spectra) and the direct possibility of quantification for some of the mass spectrometric methods can be seen as a valuable assets [1].

Analytical glow discharge mass spectrometry is a relatively young technique, profiling itself as one of the most sensitive instrumental methods for quantitative analysis of solids [2,3]. GDMS combines the stability of a glow discharge, which has a good reputation as excitation source in optical emission spectrometry [4], with the sensitivity of mass spectrometric detection. The inherent characteristics of a glow discharge provide for a uniform ionization process allowing semi-quantitative analysis without the use of standard reference materials. This feature is of special interest in the field of ultra trace analysis of high purity metals, where only very few standards are available. In this context GDMS is seen as "pretender to the throne" for older techniques like spark source mass spectrometry.

GDMS has been made commercially available by VG Instruments (now Fisons) in the early/mid eighties: the VG9000 is a high resolution double focusing instrument, which is capable of sub-ppb detection limits. The obtained mass resolution (ca. 5000) with this instrument is sufficient to overcome some of the possible interferences in a GD mass spectrum. Most of these interferences are related to the use of a working gas (usually argon); all kinds of molecular interferences due to combination of discharge gas species with the ions of major components in the sample can hamper the determination of certain (mono-isotopic) elements [5]. In 1989 VG released a low resolution quadrupole based instrument (VG Gloquad). Turner Scientific (now Finnigan) has built an interchangeable ICP/GD quadrupole mass spectrometer, with the announcement of enhanced sensitivity due to an improved ion optics system. Some research groups have worked on the development of Fourier transform mass spectrometers coupled with a glow discharge, since it became obvious that high mass resolution is almost inevitably connected with GDMS [6,7]. Results report a mass resolving power up to a few hundreds of thousands and detection limits ranging from 0.1 to 5 ppm [7]. Although optimization of these instruments towards sensitivity is necessary, this strategy seems promising. Another way of overcoming the problem of molecular interferences is the application of collision induced dissociation (CID), either in double or triple quadrupole instruments, or ion trap mass spectrometry [8-10]. Coupling of RF glow discharge devices with mass spectrometry was shown to be very useful to extend the application field of GDMS to the direct analysis of non-conducting materials [11] whereas in the conventional DC configuration this kind of samples has to be mixed with a conducting binder.

The basic physical aspects of a glow discharge are generally quite well known [12-14]. A knowledge of the plasma processes occurring in the glow discharge is unquestionably related to good analytical practice. Generally this insight can be acquired by the use of physical models. Although never complete, these models can yield valuable information if experimental data is hard or impossible to obtain. Here, the attention is focused on some fundamental aspects of a DC glow discharge, both theoretically and experimentally.

II. Operating principles of an analytical glow discharge.

A glow discharge is a simple two-electrode system placed in a gas environment at low pressure (0.1-10 torr), usually a noble gas. Applying a high enough voltage (several hundred volts) across the two electrodes will cause breakdown of the gas and a plasma will be formed in which several regions, both electrically and optically different, can be distinguished. A traditional description of the analytical glow discharge [1,4,15], where the sample serves as the cathode, takes into consideration two of these plasma regions which are of great significance regarding its application as an atomization and/or ionization source: the cathode dark space (CDS) and the negative glow (NG) (see Fig.1).

The CDS is a small region of low luminosity in front of the cathode and accounts for almost the total voltage drop across the discharge. At the anode side of the CDS, a sharp increase in light intensity to a region of bright emission marks the boundary with the negative glow. The NG is a virtually field free region and occupies in analytical GD systems most of the discharge volume. Bearing in mind the potential distribution across the discharge shown in Fig.2, the operational principles of an analytical GD can be made clear on the basis of Fig.3.

Electrons present in the CDS are quickly accelerated away from the cathode by the applied electrical field. When the energy gained is above the threshold for inelastic collisions, excitation and ionization of the (noble) gas atoms can occur, the latter causing a cascade of new (secondary) electrons. These are however not very efficient processes in the CDS; the very strong electrical field accelerates the electrons rapidly to energies beyond the level where efficient excitation and ionization takes place. This is the reason why the cathode dark space does not emit much visible radiation. Electrons entering the negative glow are now allowed to dissipate their energy to some extent because of the absence of a strong electric field. This creates a large degree of excitation and ionization of gas atoms, which in turn will relax and give the region its distinctive glow with a colour characteristic for the discharge gas. It is believed that most of the electrons injected in the negative glow will keep a beam like behaviour crossing the negative glow [16]. A lot of secondary electrons are created with much lower energies which are able to excite and ionize the gas medium with a much higher probability than the fast

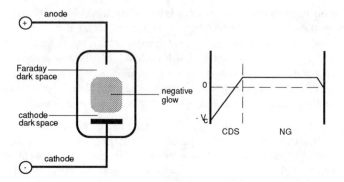

Fig. 1 :Traditional picture of a glow discharge [1].   *Fig. 2 :Potential distribution in a glow*
                                                                    discharge [14].

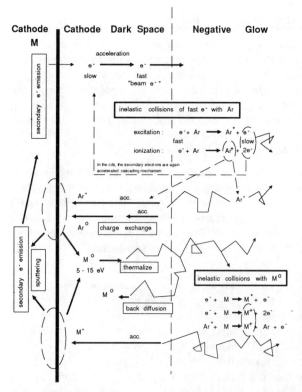

*Fig. 3 :Schematic representation of the basic processes occur-
ring in a glow discharge.*

primaries. In this way, the electron population in the negative glow is built up of three groups: a population with beam like properties in which the electrons can have energies up to the full potential drop across the cathode dark space, a population of fast secondary electrons with energies well above the excitation and ionization threshold of the gas and a slow electron group which is thermalized completely. The gas ions created by the ionization process are accelerated towards the cathode in the CDS (some ions originating from the NG will reach the NG/CDS boundary by diffusion) and cause sputter effects and secondary electron emission upon impinging on the cathode surface. Since however the environment is far from collision-less, the accelerated ions will to a large extent be subjected to charge exchange collisions. As a result, very few ions reach the cathode with energies corresponding to the full potential drop across the CDS, and energetic neutrals are formed which on their turn can bombard the cathode. It is now clear that the atomization step in a glow discharge is due to sputtering of the cathode sample by both energetic ions and neutrals. Neutral sample atoms are ejected by this process with energies ranging from a few to 15 eV [17] (secondary ion production is negligible due to the low energy of the bombarding particles). These atoms will quickly loose their energy by multiple collisions in the gas environment, whereafter their motion will be diffusion controlled. A large part diffuses towards and into the negative glow, but a considerable number will also diffuse back to the sample surface. The secondary electrons and high energy metastable gas atoms in the negative glow provide the main means for excitation and ionization of these sputtered species [18-20]. One specific feature of an analytical glow discharge can now be seen: the atomization and subsequent excitation and/or ionization steps are separated in time and space. Before analytical sampling, the species of interest (e.g. excited atoms or ions) have almost completely lost their chemical memory: the complexity of the sample matrix is converted into a normalizing gas matrix [21]. As a result GD systems are relatively free of matrix effects and a uniform sensitivity is obtained, because the formation of the analytical signal (through for instance excitation or ionization) is of non-selective nature.

A point which is up to now not completely recognized, is that cathode sample ions created in the negative glow can diffuse back to the CDS/NG boundary and then be accelerated towards the cathode surface. In this way the cathode is bombarded not only by gas ions and atoms, but also by ions of its own material. A careful mass/energy analysis of the ions which hit the cathode is carried out in our laboratory with the VG9000 mass spectrometer and the preliminary results are shown here briefly [22]. Since in the normal analytical configuration of the VG9000 for the analysis of flat samples, the analyte ions are extracted out of the negative glow, a specific glow discharge cell had to be developed to allow sampling of ions bombarding the cathode surface. This is accomplished by reversing the cell's polarity and by making the Ta-exit slit the cathode of the GD system (see Fig.4). Varying the accelerating potential at constant magnetic field and electrostatic sector voltage permits then the recording of a mass selected ion kinetic energy spectrum. Fig.5a-b demonstrates that the cathode is bombarded, among others, with low energy gas ions (argon) and high energy sample ions (Ta). With an empirical formula for the energy dependence of the sputter yield [22], the total sputter yield for these two ions could be calculated ($Y_{tot,Ar} = 0.26$, $Y_{tot,Ta} = 1.29$). Incorporating also the relative intensities of the Ar- and Ta- ion fluxes towards the cathode, it can be estimated that the sputter rate of the cathode is for 10% due to ions of its own material. For other cathode materials, this could be even more. Assuming that the argon ions originate at the NG/CDS boundary and that charge exchange collisions are the only way of dissipating energy, the theoretical energy distribution of the argon ions (dashed line in Fig.5a) differs considerably from the experimental one. Up to now no explanation is found for the almost resonant depletion of singly charged ions of low energy. The Ta-energy distribution shows a high peak at maximum (full cathode fall) energy. This means that the Ta-ions have to be formed in the negative glow and that asymmetric charge exchange with Ar atoms is a process of low cross section. Milton et al. [24] measured the mass spectrum of ions hitting the cathode sample in a glow discharge at high and low energy, and came essentially to the same conclusions. Further study is currently under way investigating the effect of the cathode material and the nature of the discharge gas.

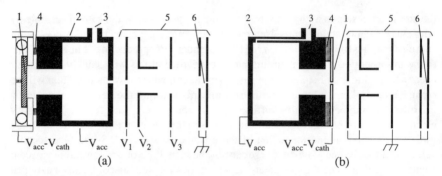

Fig. 4 :*Standard cell and focus stack configuration for the analysis of flat samples (a), and reversed configuration for the measurement of the energy distributions of ions bombarding the cathode (b); 1, cathode; 2, anode; 3, gas inlet; 4, insulator; 5, focus stack; 6, beam defining slit.*

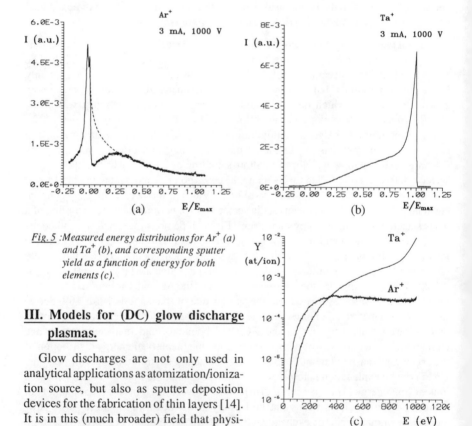

Fig. 5 :*Measured energy distributions for Ar$^+$ (a) and Ta$^+$ (b), and corresponding sputter yield as a function of energy for both elements (c).*

III. Models for (DC) glow discharge plasmas.

Glow discharges are not only used in analytical applications as atomization/ionization source, but also as sputter deposition devices for the fabrication of thin layers [14]. It is in this (much broader) field that physicists quickly came to the conclusion that modeling of GD plasmas can provide valuable information regarding the fundamental processes occurring in a glow discharge. Consequently, interesting information can be found in the applied physics literature, which is not always recognized by analytical chemists. Here we try to give a general picture of some physical models available; the list of references stated does not pretend to be complete, and is merely a starting point for those interested in more fundamental aspects of a glow discharge.

Most of the models start from a two region glow discharge, incorporating the cathode dark space (CDS) and negative glow (NG). Several approaches have been followed, with special attention for the more physical aspects of the discharge: the cathode then only acts as an electron emitter due to the bombardment with gas ions and atoms; the sputter effects and the behaviour of the sputtered cathode atoms is not of direct influence on the electrical characteristics of the discharge. One method of tackling the problem is to deal with the plasma as a (Maxwellian) fluid [25,26]. In this case continuity equations for the different species of interest (electrons and gas ions) can be set up and solved. Coupling this system with the Poisson equation for the description of the electrical field in the CDS makes the model self-consistent. Although this approach apparently gives qualitative information about the discharge, it is obvious from the electric structure of the glow discharge that it is far from equilibrium, especially in the CDS where a very strong electrical field exists. There the energy gained by the electrons from the electric field is not balanced by the energy lost by inelastic collisions. As a result, the electron energy distribution in the CDS as well as in the NG is not Maxwellian: electrons may even possess beam-like properties which become more pronounced at high discharge voltages (highly abnormal glow discharges). Electron transport can then be more accurately described by the Boltzmann transport equation [16,27-29]. The same approach can be made for the motion of ions, although it is not really necessary because these species behave more like fluids due to their limited mobility as compared to electrons. Also Monte Carlo simulations can provide an excellent way to deal with the non-equilibrium situation of the fast electrons [30] and can even be coupled with a fluid model [28] to acquire more macroscopic information about the glow discharge.

All these models intend to picture the spatial dependence of a number of discharge properties; in Fig.6 some of the obtained information is shown.

Up to now the analytical point of view has not been included. Therefore the atomization step in a GD (sputtering of the cathode) and the subsequent formation of the analytical species of interest (e.g. excited analyte atoms or analyte ions) has to be evaluated. One model dealing with the atomization step is schematically shown in Fig.7.

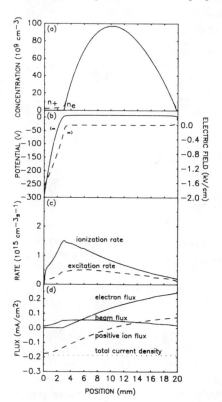

Fig.6 : *Calculated spatial variation of some discharge parameters; a, slow electron and ion densities; b, potential and electric field; c, inelastic rates, d, charged particle fluxes and total current density. Ar- discharge, 300 V, 0.6 torr [28].*

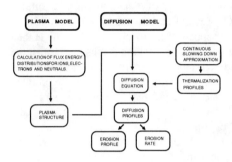

Fig. 7 :Model for glow discharge processes.

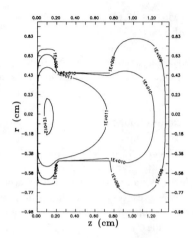

Fig. 8 :Calculated sputtered neutral atom density in a planar glow discharge configuration (depicted in Fig.4a)

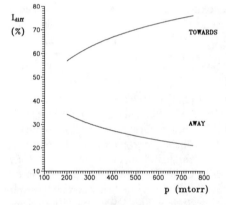

Fig. 9 :Calculated fractional sputtered particle current density towards and away from the sample surface as a function of pressure.

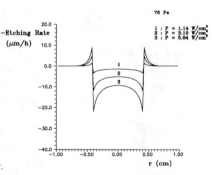

Fig.10: Theoretical crater profile for different power densities in a Mo sample after 1-h exposure to an Ar GD-plasma.

First the plasma structure is calculated by solving a set of coupled transport equations [27]. This yields the number of gas ions and fast atoms which bombard the cathode surface per second as a function of energy. With these distributions and a semi-empirical formula for the sputter yield of the cathode [23], the number of sample atoms released in the GD per second can be calculated. Since the sputtered atoms have some energy (ranging from ca. 5 to 15 eV), they have to loose this energy through collisions in the gas environment before diffusion of sputtered species can take place. This is achieved by allowing the sputtered atom flux to thermalize through a continuous slowing down approximation [31]. The result is a thermalization profile describing how the sputtered particles are distributed in the discharge after they have lost their initial energy and which can be used as a source term in the diffusion equation. Solving this diffusion equation, either in one or two dimensions, is the last step of the model and yields the sputtered atom density distribution in a certain discharge configuration (see Fig.8) [31-33]. Now it is possible to estimate the back diffusion of sputtered material back to the sample surface (Fig.9) and calculate the net etching rate of the sample. With the two-dimensional diffusion model it is possible to demonstrate that this redeposition is not uniform; this results in a convex erosion pattern of the cathode sample (see Fig.10) which is also observed experimentally. This kind of work permits to evaluate different discharge configurations (geometries) and the theoretical effect of diffusional mass transport on the analytical performance [33].

In the case of mass spectrometry, it is of course necessary to incorporate the ion formation process in the model. This has not yet been investigated; the main efforts have so far been directed towards clarifying the processes which are responsible for the ionization of sputtered material; this will be discussed in the next paragraph.

IV. Plasma diagnostics.

Physical models describing the main processes in a glow discharge plasma can provide considerable and valuable information to the analytical chemists, but should be backed up with experiments whenever possible. A number of diagnostic techniques are available for this purpose, and a few will be mentioned here.

One simple and elegant method to obtain information on GD plasma characteristics is the Langmuir probe technique [34]. It is based on the current drawn to a small probe placed in the discharge as a function of voltage applied to it, relative to the anode. If the applied voltage is sufficiently negative, only positive charge carriers (ions) will reach the probe. Conversely, if the voltage is positive, the current drawn by the probe is completely due to electrons (assuming no negative ions are formed in the plasma). In the intermediate region, only electrons with sufficient energy will be able to penetrate the electrical field surrounding the probe and be collected. In this case the probe acts as energy filter. These different zones distinguishable in the I-V curve are called ion saturation, electron saturation and electron retardation region, respectively and provide data on the plasma potential, the electron temperature and electron and ion number density [35,36]. Although simple in construction, it should be emphasized that a Langmuir probe can easily disturb the plasma and especially the size should be carefully considered: even a tiny probe can drain too much current from the discharge, and produce erroneous results.

Optical emission and absorption measurements are the most commonly used methods for gaining information regarding the number and spatial distribution of gas and sputtered species [37-39]. Also gas kinetic temperature and electron number densities can be determined spectroscopically by such studies from line profile measurements [40].

The use of a laser combined with fluorescence or absorption measurements is a very powerful diagnostic tool because of its spatial and time resolution [41,42]. Laser optogalvanic studies can give information on the electrical field and sputtered atom density variation in the cathode dark space, on the width of the dark space and on the existence of an electrical field reversal in the negative glow [42-44]. It is based on a very small perturbation of the current through the discharge due to a variation in ionization rate and/or change in ionic mobility caused by the absorption of the incoming light (optogalvanic effect). In the CDS, where a very high field exists, the possible increase in ionization is even amplified by an electron and ion avalanche: electrons created by the laser light are quickly accelerated by the field and cause additional ionization. Other optogalvanic studies concern the NG, where a laser of suitable frequency can selectively excite gas ions to energy states which have lower mobilities; in this case a small decrease in discharge current is anticipated [45]. The application of laser diagnostics has a great advantage compared to Langmuir probe techniques or electron beam deflection methods because of its non-disturbing nature.

Mass spectrometry can also be used for characterizing glow discharge processes. It is interesting to note that the first generation of GDMS instruments was not developed to do analytical work, but served merely as diagnostic tool [46-49]. Later on, also its analytical capabilities became clear, and now the situation is somewhat reversed.

Whereas the attention of the physicists used to be focused on the (noble) gas ionic species, the ionization of analyte atoms is of course of more interest for the analytical chemist. It is generally accepted that ionization of sputtered sample atoms in a glow discharge is mainly caused by two processes, electron impact and Penning ionization. Many other ionization routes are possible, but of much lower probability [1,20]. Penning ionization is the result of a collision between a metastable gas atom with an atomic species of which the ionization energy is below the energy content stored in the excited metastable state of the gas atom. It is a non-selective process since this energy content (e.g. Ar: 11.57 and 11.78 eV) exceeds the ionization potential of most elements, which is one of the reasons for the more or less uniform sensitivity obtained in GDMS. A lot of effort has been put in the evaluation of the relative importance of Penning ionization compared to electron impact ionization. Since Penning ionization is strongly related to the number density of metastable species, the population or depopulation of these energy levels results in a variation of the ionization efficiency and thus also in the ion signals. Depopulation can be achieved by introducing a quenching gas (e.g. nitrogen or methane) [20,50,51]. The application of a tunable laser allows both a selective population and depopulation of metastable energy levels. The latter is achieved by exciting the metastable gas atom to a nonmetastable state making radiative decay possible [45]. All these studies reveal that the Penning process plays a major role in the ionization of both the sputtered sample atoms and the gas atoms. Although most investigations are qualitative, it is believed that ionization of sputtered species in a glow discharge maintained at pressures below 1 torr is for 40 to 80 % due to Penning ionization, depending on the discharge characteristics (current and pressure) [51].

References

[1]   W.W. Harrison, *Glow discharge mass spectrometry*, in F. Adams, R. Gijbels and R. Van Grieken (Eds), *Inorganic Mass Spectrometry*, Wiley, New York (1988).

[2]   N. Sanderson, E. Hall, J. Clark, P. Charalambous and D. Hall, Mikrochim. Acta, **1**, 275 (1987).

[3]   R. Gijbels, Talanta, **37**, 363 (1990).

[4]   S. Caroli, J. Anal. Atom. Spectrom., **2**, 661 (1987).

[5]   A. Raith, W. Vieth, J.C. Huneke and R.C. Hutton, J. Anal. Atom. Spectrom., **7**, 943 (1992).

[6]   R.K. Marcus, P.R. Cable, D.C. Duckworth, M.V. Buchanan, J.M. Pochkowski and R.R. Weller, Appl. Spectrosc., **46**, 1327 (1992).

[7]   C.M. Barshick and J.R. Eyler, J. Am. Soc. Mass Spectrom., **3**, 122 (1992).

[8]   D. Duckworth and R.K. Marcus, Appl. Spectrosc., **44**, 649 (1990).

[9]   F. King, A. McCormack and W.W. Harrison, J. Anal. Atom. Spectrom., **3**, 883 (1988).

[10] S. McLuckey, G. Glish, D. Duckworth and R.K. Marcus, Anal. Chem., **64**, 1606 (1992).

[11] D. Duckworth and R.K. Marcus, Anal. Chem., **61**, 1879 (1989).

[12] M.J. Druyvesteyn and F.M. Penning, Rev. Mod. Phys., **12**, 87 (1940).

[13] G. Francis, in S. Flügge (Ed.), *Handbuch der Physik*, vol. 22, Springer-Verlag, Berlin (1956).

[14] B. Chapman, *Glow Discharge Processes*, Wiley, New York (1980).

[15] W.W. Harrison and B.L. Bentz, Prog.Analyt. Spectrosc., **11**, 53 (1988).

[16] R.J. Carman and A. Maitland, J. Phys. D: Appl. Phys., **20**, 1021 (1987).

[17] R.V. Stuart, G.K. Wehner and G.S. Anderson, J. Appl. Phys., **40**, 803 (1969).

[18] R.L. Smith, D. Serxner and K.R. Hess, Anal. Chem., **61**, 1103 (1989).

[19] W.Vieth and J.C. Huneke, Spectrochim. Acta, **45B**, 941 (1990).

[20] M.K. Levy, D. Serxner, A.D. Angstadt, R.L. Smith and K.R. Hess, Spectrochim. Acta, **46B**, 253 (1991).

[21] W.W. Harrison, J. Anal. Atom. Spectrom., **7**, 75 (1992).

[22] M. van Straaten, unpublished results.

[23] Y. Yamamura, N. Matsunami and N. Itoh, Radiat. Eff., **71**, 65 (1983).

[24] D. Milton and R. Mason, 12th International Mass Spectrometry Conference, 26-30August 1991, Amsterdam, Book of Abstracts, p 250.

[25] S. Pirooz, P.A. Ramachandran and B. Abraham-Shrauner, IEEE Trans. Plasma Sci., **19**, 408 (1991).

[26] J.-P. Boeuf, J. Appl. Phys., **63**, 1342 (1988).

[27] I. Abril, Comp. Phys. Comm., **51**, 413 (1988).

[28] M. Surendra, D.B. Graves and G.M. Jellum, Phys. Rev. A, **41**, 1112 (1990).

[29] T.J. Sommerer, W.N.G. Hitchon and J.E. Lawler, Phys. Rev. A, **39**, 6356 (1989).

[30] J.P. Boeuf and E. Marode, J. Phys D: Appl. Phys., **15**, 2169 (1982).

[31] J.A. Valles-Abarca and A. Gras-Marti, J. Appl. Phys., **55**, 1370 (1984).

[32] M. van Straaten, A. Vertes and R. Gijbels, Spectrochim. Acta, **46B**, 281 (1991).

[33] M. van Straaten, R. Gijbels and A. Vertes, Anal. Chem., **64**, 1855 (1992).

[34] K. Hoefler, *Plasmadiagnostik bei plasmaunterstützten Dünnschicht-techniken*, Firmenschrift Balzers, Liechtenstein, BG 800 184PD (8402).

[35] D. Fang and R.K. Marcus, Spectrochim. Acta, **45B**, 1053 (1990).

[36] N. Jakubowski, D. Stuewer and G. Toelg, Int. J. Mass Spectrom. Ion Proc., **71**, 183 (1986).

[37] T.J. Loving and W.W. Harrison, Anal. Chem., **55**, 1523 (1983).

[38] N.P. Ferreira and H.G.C. Human, Spectrochim. Acta, **36B**, 215 (1981).

[39] A.J. Stirling and W.D. Westwood, J. Phys. D: Appl. Phys., **4**, 246 (1971).

[40] N.P. Ferreira, H.G.C. Human and L.R.P. Butler, Spectrochim. Acta, **35B**, 287 (1980).

[41] C. Van Dijk, B.W. Smith and J.D. Winefordner, Spectrochim. Acta, **37B**, 759 (1982).

[42] E.A. Den Hartog, D.A. Doughty and J.E. Lawler, Phys. Rev. A, **38**, 2471 (1988).

[43] F. Babin and J.-M. Gagné, Appl. Phys. B, **54**, 35 (1992).

[44] R.A. Gottscho, A. Mitchell, G.R. Scheller and Y.Y. Chan, Phys. Rev. A, **40**, 6407 (1989).

[45] K.R. Hess and W.W. Harrison, Anal. Chem., **60**, 691 (1988).

[46] P.F. Knewstubb and A.W. Tickner, J. Chem. Phys., **36**, 674 (1962).

[47] P.F. Knewstubb and A.W. Tickner, J. Chem. Phys., **36**, 684 (1962).

[48] M. Pahl and U. Weimer, Z. Naturforsch., **12a**, 926 (1957).

[49] W.D. Davis and T.A. Vanderslice, Phys. Rev., **131**, 219 (1963).

[50] K. Wagatsuma and K. Hirokawa, Anal. Chem., **61**, 326 (1989(.

[51] R. L. Smith, D. Serxner and K.R. Hess, Anal. Chem., **61**, 1103 (1989).

Depth Profiling Study of Scale Formed on High Si Content Steels Using GDMS

M. Pichilingi and R. S. Mason
DEPARTMENT OF CHEMISTRY, UNIVERSITY COLLEGE SWANSEA, SINGLETON PARK, SWANSEA SA2 8PP, UK

D. Gilmour
VG ELEMENTAL (FISONS INSTRUMENTS), ION PATH, ROAD 3, WINSFORD, CHESHIRE, UK

N. Croall, M. Westacott, and D. C. Richards
BRITISH STEEL TECHNICAL, WELSH LAB., PORT TALBOT, WEST GLAMORGAN, UK

1 INTRODUCTION

GDMS is now well accepted as a good technique for elemental trace analysis in bulk metals[1]. Its use as a surface analytical technique is less well developed[2]. Most surface analysis techniques are restricted to probing depths at or very close to the surface, and focus usually on very small spot sizes at any one time with the need to scan the surface. In the steel industry, where the surface composition of steel plate is often crucial to its function and corrosion properties, there is often a need to probe *near* surface layers to depths of many microns. In addition, after rolling and annealing etc., the composition of the steel surface may vary significantly over different parts of the surface. Glow Discharge methods have a distinct advantage in this respect, since they are capable of penetrating to micron depths, quickly, over a relatively large surface area, averaging out the often considerable variations in the detailed composition of the surface. GD Atomic Emission Spectroscopy has been used over many years for this reason[3].

Problems in processing steel often arise from the presence of "scale", particularly in the cold rolling stage, where its presence can cause excessive wear on the rollers. Scale is the surface layer formed by rapid oxidation when hot steel cools in air, and is therefore composed mainly of the various oxides of iron, depending mainly on the temperature of formation and cooling rate, but also on other factors such as cracks in the scale. The bulk of the scale is usually wustite (75-77% iron). The thick hard brittle layers can be removed prior to further processing by mechanical means, but the thinner more persistent layers must be removed by 'pickling' in acid. In this study we were investigating the use GDMS to directly probe the composition of the different layers of scale as an aid to diagnosis of technical problems further down the line during cold rolling of the steel. This paper describes its successful application to a specific problem associated with high Si content steel.

There are two types of high silicon content (~3%) steel manufactured for electrical use, transil and unisil[4] (transil differs from unisil in that it contains added aluminium) both of which are more difficult to pickle than normal, requiring HF in the pickling liquor. In experiments to test the use of alternative acid mixtures, although the scale was removed problems were encountered with a final resistant layer. Reflectance measurements by FTIR showed Si-O absorption, which increased with pickling time. It was not known if this increase occurred because the acid removed the iron oxides preferentially causing enrichment of Si_xO_y or if there was a truly Si rich layer at lower levels. GDMS was used in combination with optical microscopy to study these layers.

2 EXPERIMENTAL METHOD AND INSTRUMENTATION

The instrument used was the "Gloquad" (Fisons, VG Elemental)[5], using the standard source and inlet system, designed for both bulk analysis and depth profiling.

Samples of steel strip, previously pickled to various degrees, were subjected to an Argon discharge, under 2 sets of conditions giving two different erosion rates.

The faster conditions (II) were employed for analyses down to depths of 30 μ, whilst the slower conditions (I) penetrated through no more than 4μ. The depths were determined by discharging a sample for 60 minutes or more, measuring the crater depth by a "Talysurf" (Rank, Taylor and Hobson) and assuming a constant erosion rate for similar samples.

Table 1: Discharge parameters

	CONDITIONS	EROSION RATE (μ/minute)
I	0.8 kV, 0.2 ma	0.14
II	1.3 kV, 0.2 ma	0.25

Area exposed : 4mm diameter; $1 \mu = 1 \times 10^{-6}$ m

The elements of interest in this study were: C (**12**), O (**16**), Al (**27**), Si (28, **29 and 30**), O_2 (**32**), S (32 and **34**) and Fe (**56** and 54), where the figures in bold are the mass to charge ratios selected for monitoring, being chosen in an attempt to avoid likely isobaric interferences. Since the relative abundances were not required with good precision, no attempt was made to find relative sensitivity factors (RSF) for these samples.

3 RESULTS AND DISCUSSION

Micrographs for typical samples of unisil and transil, showing sections through the scale and underlying steel substrate are shown in figure 1. Unisil (A) has a layer of scale of widely varying thicknesses over the surface (2-40 μ). A distinct inner layer,approximately 2 μ thick, can be identified. Transil (B) shows a layer of 4-15 μ thickness, thinner than unisil but an inner layer which is significantly thicker (8-10 μ).

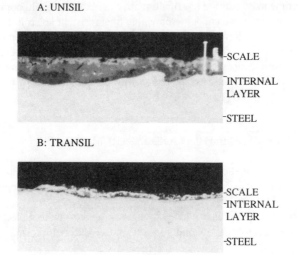

A: UNISIL

-SCALE

INTERNAL LAYER

-STEEL

B: TRANSIL

-SCALE
-INTERNAL LAYER

-STEEL

<u>Figure 1</u>; Micrographs of sections through the scale of samples of **Unisil** and **Transil**.

This layer can only just be seen here in figure 1B (though it is more obvious in the original micrographs) as a very light grey region.

Depth profiles were obtained for a number of examples of each type of sample. Profiles down to 30 μ are shown in figures 2 and 3. *These are not RSF corrected.* Nevertheless the raw data still gives percentage abundances for Si and Al in the bulk which are close to their actual values. The O levels within the scale are high as expected but a factor of 10 lower than the known percentage levels. This is at least partly because the ionisation potential of O is very high by comparison with metals, which will give it a very low relative sensitivity factor. A second reason may be that a fair proportion of the signal is locked into the molecular ions (see figure 7), which were not monitored on this occasion (though in retrospect this might have been valuable). It would be difficult in any case to find suitable standards for determining RSFs in this varying matrix. However this does not detract from the diagnostic value of the information obtained.

The interfacial regions showup quite distinctly as "hiccups" in the signal, particularly m/z = 29. The Unisil sample (fig. 2) shows interfacial regions at 3 and 6 μ respectively, whilst in Transil(fig. 3) they occur at 2 and 15 μ. These regions correlate well with the scale and internal layers seen on the micrographs. The scale layer appears thinner, but it is very likely that the erosion rate of scale layers, which consists of magnetite (Fe_3O_4) and wustite (Fe:75-77%, O:25-23%), is lower than

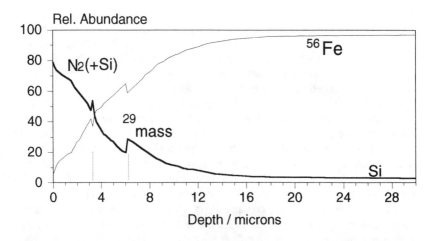

Figure 2   **Unisil**; depth profiles of m/z = 56 and 29; discharge conditions II

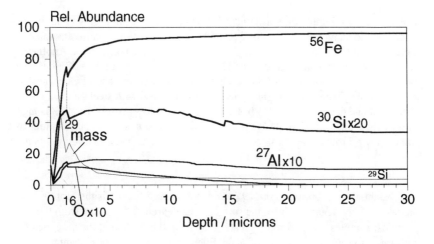

Figure 3 **Transil**; depth profiles showing m/z = 16,27,29,30 and 56; disch. param. II

bulk steel. This is supported by our experiments to measure sputtering rates of iron in the glow discharge, when even oxide layers formed on freshly exposed iron surfaces are found to significantly inhibit the sputtering rate. No attempt was made here to study the molecular iron oxides as such, some of which are sputtered intact as molecular species into the plasma, and are quite clearly seen as complete ions within the spectrum (see figure 7).

Although ^{28}Si is by far the most abundant isotope, it can not be used to monitor Si because of potential isobaric overlap with mainly N_2, but also possibly C_2H_4, CO and Fe^{++}. The signal at mass 29 settles to approx. 3% of the iron (mass 56) signal, which is equivalent to the bulk Si content in the steel, however it does not overlap with the mass 30 (^{30}Si) signal early on in the discharge. Mass 29 is actually completely dominant relative to 56 (Fe) at the start. $C_2H_5^+$ is often present in GD spectra resulting from organic contamination. This would be present at the start if for example mineral oil is trapped within the scale, but this is not likely because the steel comes into contact with oil only at the cold rolling stage. To check we can compare the profile for ^{12}C, which is expected to follow 29 in its behaviour if the source is hydrocarbon. This is shown by the depth profiles of steel strip known to be contaminated with oil (see figure 4), in this case with much lower levels of bulk Si and after full processing, hence without scale. Again the rapidly changing mass 29 signal is very high relative to iron but the m/z =12 signal due to C follows 29 very closely at about 10% of the abundance.

The profiles of a sample of unisil under the less aggressive sputtering conditions (figure 5), do show the profiles for ^{12}C and ^{29}M to be distinctly different and therefore that these samples are *not* significantly contaminated with oil. Another possible interference at mass 29 is N_2H^+ . Peaks due to N_2 at m/z = 28 (identified by accurate mass measurement on a high resolution machine), and of CO_2 (peaks at m/z = 44 (CO_2^+) and 45 (CO_2H^+)) are always observed with steel samples as a result of degassing. Although N_2 has a comparatively low proton affinity it is higher than Ar, and since ArH^+ is always present in the plasma in significant quantities this will readily protonate N_2 (exothermic proton transfer occurs at collision frequency)

$$ArH^+ + N_2 \xrightarrow{\Delta H = -123kJ \ mol^{-1} \ (ref.\ 6)} Ar + N_2H^+$$

 A substantial contribution to the 29 peak at the start of the run is therefore simply the result of N_2 degassing from the surface. The degree of degassing from these particular samples is significantly greater than from the surface of the base metal, because of the spongy structure of scale. Even so it becomes insignificant after 300-400 s into the run and reduces to a level which is consistent with the Si levels known to be present in the sample. After this time the 29 signal is representative of Si present in the discharge.

The profiles in figure 2-4 show that the transil scale is enriched in both Al and Si by approximately 0.2 and 1 % respectively, compared to the bulk levels at 0.6 and 3-3.5 %. The abundance of oxygen (O) sputtered off from these layers is very high compared to normal and bulk levels as would be expected for material which has a high iron oxide content.

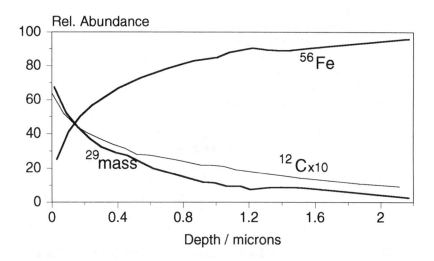

Figure 4; 'Depth' profile of m/z = 12, 29 and 56, from steel surface contaminated with oil; disch. param. II.

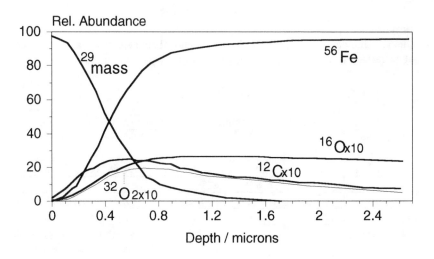

Figure 5; **Unisil**; depth profiles through the first 3μ of scale; disch. param. I

The elemental changes observed in crossing one layer to the next are not dramatic, however the interfacial regions can be discerned by small, abrupt, but nevertheless significant temporary changes in composition. Real effects over very short distances are bound to be reduced because of the convoluting effect of the considerable degree of back-diffusion which occurs at the relatively high pressures of this technique. However the crossover points are sharply revealed by the degassing component of the 29 peak which does increase significantly in abundance and sharply at depths equivalent to the boundaries. This is presumably because there are likely to be more cavities releasing gas in these interfacial regions. In addition the gas would not resettle back onto the surface as occurs for the refractory elements.

Figures 5 and 6 show profiles for unisil and transil respectively, obtained under the less agressive sputter conditions, which therefore penetrate only a few microns into the scale layer.

Unisil
As before degassing is responsible for the earlier part of the 29 profile, which however decays to very low levels (<1%) suggesting that the scale is depleted in Si at these levels close to the surface of the scale. The O abundance is high as expected. An interesting observation is the peak for O_2 for which there is a delay before it is desorbed. This is almost certainly due to a layer of chemisorbed O_2 which usually sits on oxide layers as O_2^-. C also shows a profile characteristic of desorption for a chemically adsorbed species. This could be from CO_2. Unfortunately the 44 peak was not monitored here.

Transil
In the very top part (<1 μ) of the scale both Si and Al are depleted relative to their bulk levels, but increase to bulk level or slightly above at ~4μ.

Profiles of Surface Etched Transil

Table 2: Surface Elemental Composition[#] Variation with 'Pickling' Time

'pickling' time /s	Si	Al	O	Micrograph Evidence
0[*]	3.5	0.6	<0.1	2 scale layers visible
	4.8	0.9		
30	3.5	1.0	0.2	top layer removed
60	11	1.0	1.1	internal layer still visible
120	11	1.2	1.2	as above, but starting to breach interface to metal
240	24	3.0	3.8	both layers apparently removed
330	3	0.6	<0.1	base metal exposed

\# These concentrations are taken 1000 s after the onset of discharge(I), equivalent to a penetration of ~2μ into the surface.

* bulk levels taken from the depth profile ~30 μ down from the surface.

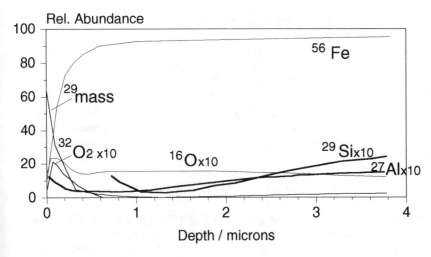

Figure 6; **Transil**; depth profiles through the first 4μ of scale; disch param. II

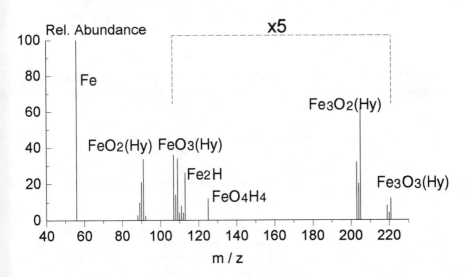

Figure 7 ; GD spectrum of an iron cathode previously exposed to air.

The scale surface of samples of transil steel were etched for varying times in 24% H_2SO_4. The pertinent data are summarised in table 2.

After pickling for only 30 s, micrograph evidence showed that the iron oxide layer had been removed leaving the white/ light grey inner layer exposed. The depth profile of this layer is shown in figure 7. Initially there is an expulsion of O_2, S, Al and O from the top of the layer, but levels settle down to give Si and Al levels at 3.5 and 0.7 % more or less as normal, and very similar to the full depth profile showing this inner layer in figure 3. However as the sample is pickled for progressively longer times the Si and O content in the surface layer increases very significantly up to a value of 26%, until a critical value is reached when the value suddenly reduces to the normal bulk values. These results are consistent with IR measurements which show increasing absorption due to the Si-O bond as pickling time is increased.

These data contrast sharply with the full depth profile through the scale and into the base metal which shows that Si enrichment within the scale only occurs to the extent of ~1% above bulk levels. This shows therefore that the high Si enrichment in the acid etched samples is caused by the preferential reaction of the acid with the iron oxides, leaving Silica behind on the surface as a fine thin layer, and to a much lesser extent alumina. Even when the micrograph evidence appears to show base metal has been reached, after 240 s in the pickling tank, the GD measurements shows that the Si level is at its maximum! Eventually (after 330 s) however base metal is reached, the silica and alumina presumably being undermined and dropping off the surface into the pickling tank, so that Si, Al and O reduce back to normal bulk levels.

4 CONCLUSIONS

This study highlights a potential problem in depth profiling the innate surfaces of samples of steel by GDMS. There is initially a high degree of interference from degassing of the sample, which will particularly affect the low mass ranges. The occurrence of N_2H^+ at such relatively high levels, for example, was unexpected. This can be overcome in a low resolution instrument provided there is a suitable elemental isotope with no isobaric overlap. A significant improvement to the technique, at least from the point of view of steel depth profile analysis, might be obtained by the provision of a facility for thermal degassing of the sample in vacuum, prior to discharge.

The use of micrographs has proved invaluable, in this case, to the interpretation of the depth profiling data. The different layers are picked out quite clearly by GDMS, not because of dramatic changes in elemental concentrations, but because of significant increases in the rates of degassing at the interfaces. The averaged concentrations of Si and Al are enriched within the scale, but only by about 1%. These layers do show significant changes in the amount of O present, decreasing steeply across the inner layer and into the base metal. The scale on these samples are therefore still composed mainly of iron oxides, whose elemental composition remains more or less constant but differing in structure between the outer and inner layers.

The data are fully consistent with what is already known of the composition of scale on steel. However the amount of Si and Al present is enough to create a significant problem during the pickling stage of these high Silicon steels. This study shows that acid reaction leaves behind fine thin layers of Silica and, to a much lesser extent, Alumina over the surface. It is not due to enrichment of these species at the surface of the bulk metal caused by previous heat treatments. These are hard materials which cause excessive wear on the rollers during cold rolling.

This case study illustrates the potential diagnostic value of depth profiling of steel by GDMS, especially when used in conjunction with micrograph evidence.

REFERENCES

1. P. M. Charalambous, Steel Research, 1987, 58, 197.
2. D. J. Hall and N. E. Sanderson, Surface and Interface Analysis, 1988, 11, 40.
3. D. C. Richards, PhD Thesis, University of Swansea, 1991.
4. These are tradenames of British Steel plc.
5. A. Raith and R. C. Morton, J. Anal. and Atomic Spectr., 1992, 7, 623.
6. Data taken from S. G. Lias, J. F. Liebmann and R. D. Levin, J. Phys. Chem. Ref. Data, 1984, 13, 695.

Application of ICP-MS to Steel, Other Metals, and Metal Alloys

H. Ekstroem and I. Gustavsson
SWEDISH INSTITUTE FOR METALS RESEARCH, DROTTNING KRISTINAS VAEG 48, S 114 28 STOCKHOLM, SWEDEN

1 INTRODUCTION

Steel and metals consist of one dominating element surplus micro and trace elements. Metal alloys and stainless steel contain one or two extra major constituents besides the main matrix element. To carry out analysis of all these metallurgical samples using ICP-MS can be troublesome since the major ion or ions can interfere severely at the determination of trace elements. Suppression of the signal of the analyte ions might happen. Therefore an investigation was performed in our laboratory in order to evaluate the influence of the matrix ions on the detector signal of some trace elements.

Regularly, certified reference materials (CRMs) are analysed in order to check the accuracy. Several metallurgical CRMs are available, and in this study the figures for steel and zirconium are reported.

An ICP-MS method for phosphorus determination has been investigated with the purpose of obtaining a fast analytical method and also of getting a complement to the tedious spectrophotometric method commonly used.[1]

2 EXPERIMENTAL

The measurements were carried out with an inductively coupled plasma mass spectrometer (ICP-MS), VG Elemental Plasma Quad 2 Plus (Winsford, UK), equipped with a Gilson sample exchanger.

3 MATRIX EFFECTS

The difference in concentration between a trace element and a major matrix ion can be huge when analysing metallurgical samples, six to seven orders of magnitude. Matrix effects have been reported for other types of samples.[2] It was found to be of importance to carry out such studies on the common metallurgical elements Fe, Cu, Ni, Zr and W. The matrix effects for a large number of analyte ions were investigated at various nebulizer gas flow rates, as it has been reported that the gas flow rate might influence the matrix effects. The concentration of each analyte ion was kept at a concentration of 5 ng/ml throughout the investigation.
Some typical results are given for In in Figure 1.

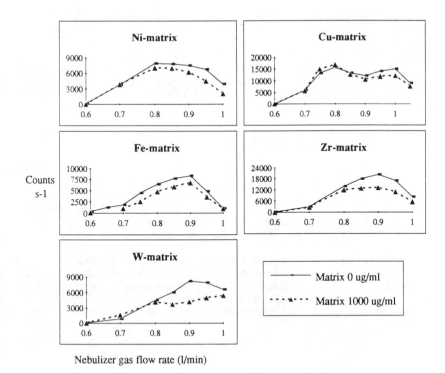

Nebulizer gas flow rate (l/min)

<u>Figure 1</u> The influences of Ni, Cu, Fe, Zr and W matrices (1000 mg/ml) on the signal of the analyte ion In (5 ng/ml).

The suppression effects varied from 0-50% and the matrices could be ranked with increasing matrix interferences on the analytes as

$$Cu < Ni < Fe < Zr < W$$

As could be expected the low mass number elements were more affected than the heavier ones. In all matrices the boron signal was heavily decreased, e.g. in the Ni-matrix to 30%. The heavy W-matrix even suppressed the U-signal to about 50% compared to solutions without the W-matrix. See Figure 2.

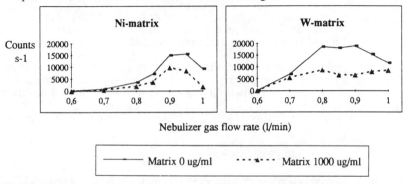

Nebulizer gas flow rate (l/min)

<u>Figure 2</u> Influence of the Ni and W matrices on the signals of B and U (both 5 ng/ml), respectively.

A broad investigation of the influence of the matrix ions Fe, Ni, Cu, Zr and W is given in a technical report.[3]

Furthermore it can be observed that the suppression effects were less at lower nebulizer gas flow rates. The internal standard used could then compensate for some of the suppression effects besides for any drift in the analytical signal. In conclusion, matrix matched standards and internal standardisation are of utmost importance to correct for matrix suppression effects, when analysing metallurgical samples.

4 MULTIELEMENT MATRIX

Stainless steel and other high alloy steel materials consist of at least two matrix ions. Besides the iron, the two most common ones are chromium and nickel. In view of the fact that the matrix composition is not always exactly known, the preparation of matched standards is a laborious procedure. If chromium and nickel behave in the same way as iron does, considering the suppression effects, it should be sufficient to use standards, which are only matched with iron.

Several trace elements at a concentration of 10 ng/ml were determined in acidified solutions matched with the three elements Fe, Ni and Cr at various concentrations. Subtractions with matrix matched blanks were made to avoid a change in signal because of contamination originating from the matching materials. The combinations investigated can be followed in Figure 3. The study will be published in a future report.[4]

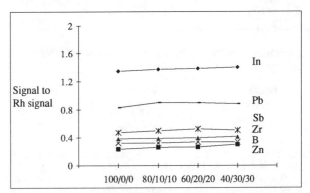

Figure 3   The change in signal for the elements B, Zn, Zr, Sb, In and Pb, each element kept at a concentration of 10 ng/ml, when the matrix elements Fe/Ni/Cr are altered in % from 100/0/0 to 40/30/30.

It could be concluded that if the nebulizer gas flow rate is kept slightly lower than that giving a maximum analytical signal and also by introducing internal standard correction, the influence of the two alloying elements on the analytical signal could be neglected. This means that trace elements in steel can be determined by the usage of iron matched standards. Hence a time saving analytical procedure is achieved.

5 IRON MATRIX

The most common procedure to dissolve steel samples is the usage of an acid mixture of HCl/HNO_3. If HCl is added, the metallurgically important elements V, As and Cr cannot be determined because of severe interferences.[5] An alternative dissolution procedure is to replace HCl with HF and then the three elements can be determined. The drawback by dissolving with HF is the fact that glassware can no longer be used. In order to avoid the etching of glassware in the ICP-MS instrument, a Teflon spray chamber equipped with a V-Groove de Galan nebulizer was installed, and furthermore the Ni sample cone was replaced by a Pt sample cone. Determination of the element boron in metallurgical samples demands a time consuming method, but the ICP-MS technique opens a new possibility. When installing the Teflon introduction system the background contamination of boron from the glass can be minimized and thus facilitate the determination of boron at trace levels.

Table 1   ICP-MS analyses - Results obtained from the analyses of the CRMs JSS 003-3 and JK 1C. All values are given in µg/g.

Sample	JSS 003-3		
Element	Average	SD	Cert.[1]
Mg	0.62	0.52	(<1;<0.1;0.6)
Ti	0.69	0.04	(0.1-0.6)
V	0.18	0.07	(0.1;<1;<0.1)
Mn	47.9	0.4	48
Co	9.0	0.2	10
Ni	8.2	0.8	8
Cu	13.0	0.7	14
Zn	<0.3		(0.4;0.5)
As	0.87	0.25	(0.8;<1;<0.1)
Nb	0.04	0.03	(0.1;<1;<0.3)
Mo	0.22	0.05	(0.5;1;2;<1)

[1] The values within the brackets are values reported by different laboratories.

Table 2   ICP-MS analyses - Results obtained from the analyses of the JK 1C. All values are given in µg/g.

Sample	JK 1C		
Element	Average	SD	Cert.
B	1.6	0.4	[1](2)
Al	9.1	0.7	(<20)
Ti	0.40	0.06	(0.1)
V	5.2	0.2	5
Cr	12	0.5	12
Mn	60	1.0	59
Co	46	1.0	46
Ni	88	2.0	90
Cu	14	0.4	14
Zn	3.2	0.9	2
As	5.3	0.5	(4;11;5.1)
Nb	0.19	0.005	(<0.1;1)
Mo	2.3	0.5	3
Sn	28	1.1	24
Sb	2.1	0.09	(2.1)
Ta	0.06	0.01	(<0.001)
Pb	0.13	0.03	(<2;<1)
Bi	0.02	0.02	(<1)

[1] The values within the brackets are values reported by different laboratories.

The analysis of two certified reference materials (JSS 003-3 and JK 1C) were performed using the peak jumping mode measuring five points per peak for the acquisition of data. Indium and rhodium were used as internal standards. The results obtained at the analysis of the certified reference materials are presented in Table 1. The concentrations are given in µg/g (ppm) which means that a back calculation to the original sample has been made from the solution. The dilution factor is 1000 times.

6 ZIRCONIUM MATRIX

The accuracy of the zirconium analysis using ICP-MS for trace elements were investigated by analysing the CRMs JAERI Z11 and NIST SRM 1239.

Sample Preparation

About 100 mg of the sample was dissolved in 2 ml HF and 0.2 ml H_2O_2 and made up to 100 ml volume. Rhodium and iridium were added as internal standards. Indium as internal standard should be avoided when analysing Zr, since Zr alloys often contain high concentrations of Sn, which overlaps the In-isotope at 115 amu. Scandium (^{45}Sc) as internal standard should also be avoided because of the spectral interference from $^{90}Zr^{2+}$.

The analysis took place by employing the Teflon introduction system and the results obtained, Table 3, agreed well with reference values.

Problems have arisen during the determination of Cd in Zr-materials, which is an element of interest in the Zr-metallurgy. Most of the Zr-materials contain Mo and since MoO and MoOH are species easily formed, Figure 3 and 4, the oxides and hydroxides of the different Mo-isotopes will overlap the Cd-isotopes in the mass spectra. Cd-isotopes are also overlapped by the oxides of Zr. This contributes to a very high background level for the Cd-isotopes and will make it extremely difficult to determine Cd at low levels in Zr-materials.

Many steels contain Mo, and consequently the determination of Cd can then be difficult to carry out. In addition there are also difficulties to determine ^{128}Te and ^{130}Te as there are isobaric overlaps from $^{96}MoO_2^+$ and $^{95}MoCl^+$, respectively.

Table 3   ICP-MS analyses - Results obtaind from the analyses of the Zr reference materials NIST SRM 1239 and JAERI Z11. The results are given in µg/g.

Sample	NIST SRM 1239			JAERI Z11		
Element	Average	SD	Cert.	Average	SD	Cert.
B	0.22	0.006	(0.25)	1.0	0.09	1.1
Mg	0.46	0.15		0.71	0.18	
Mn	48.7	3.2	(50)	4.9	0.3	5
Co	15.8	0.9	(15)	6.7	1.5	6
Cu	35.0	2.2	(30)	40.0	1.4	40
Hf	80.3	1.7	77	72.2	1.0	71
Ta	382	8	(400)	0.73	0.04	
W	44.9	1.0	(45)	13.6	2.0	13
Pb	38.6	0.4	(30)	11.6	1.5	12
U	0.84	0.02		0.87	0.05	0.8

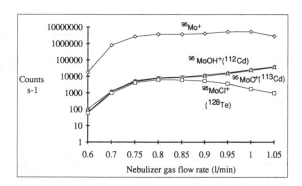

Figure 4 The formation of ^{95}Mo molecular ions in the ICP-MS when analysing a solution containing 30 µg/ml Mo.

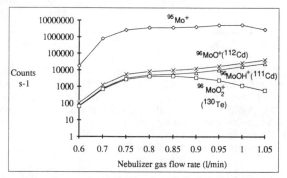

Figure 5 The formation of ^{96}Mo molecular ions in the ICP-MS when analysing a solution containing 30 µg/ml Mo.

7 LIMIT OF DETECTION

Ten measurements were performed both on the blank and on the multielement standard with an analyte concentration of 5 ng/ml. The blank and the standard were matched with 1000 µg/ml of the matrix investigated. Since the procedure for normal analysis comprises a blank subtraction, this was made.

The limit of detection (3σ) for some elements in the matrices Fe and Zr are given in Table 4.

8 DETERMINATION OF PHOSPHORUS

Phosphorus determination using ICP-MS can be troublesome since the monoisotopic phosphorous (^{31}P) has two adjacent interfering molecular ions in the mass spectra originating from NO^+ (mass number 30) and O_2^+ (mass number 32), respectively. The resolution has to be optimized carefully in order to separate two big peaks from a small phosphorus peak.

Table 4 Limit of detection (LOD) for some elements in Fe and Zr materials, respectively. (The values are given in $\mu g/g$).

Fe-matrix				Zr-matrix	
Element	LOD	Element	LOD	Element	LOD
Be	0.08	Mo	0.5	Be	0.11
B	0.6	In	0.01	B	1.0
Mg	0.2	Sb	0.05	Mn	0.6
Sc	0.2	Ir	0.03	Co	0.2
Ti	0.1	Pb	0.06	Cu	0.5
Cr	2.0	Bi	0.01	Mo	0.2
Mn	0.8			Rh	0.04
Co	1.3			W	0.08
Ni	0.4			Ir	0.08
Cu	0.5			Pb	0.08
Zn	0.6			Th	0.02
Zr	1.5			U	0.1

Sample preparation

The sample, 100 mg, was dissolved in a 5 ml acid mixture of $HClO_4/HNO_3/H_2O$ (2/1/1) with a second addition of 2 ml HCl. The oxidation potential had to be kept high to avoid losses of phosphorus as phosphin. The internal standard used was beryllium and the sample was made up to 100 ml volume with ultrapure water. The phosphorus concentration was evaluated by means of a matched standard calibration curve performing the blank subtraction technique. Two CRMs, both low alloy steels, were analysed and the results are given in Table 5.

Table 5 The concentration of phosphorus in two low alloy steel CRMs using the ICP-MS technique. Number of samples analysed=n.

CRM	Concentration determined ($\mu g/g$)	SD ($\mu g/g$)	n	Certified value ($\mu g/g$)
JK 2D	77	4	4	78
JK 21	148	1	5	148

The results obtained coincided well with certified values. The detection limit was 10 $\mu g/g$, which means that most low alloy steels can be determined by the ICP-MS method, which offers a fast analytical method.

9 CONCLUSIONS

Matrix ions suppress the detector signal of the trace analyte ions and it is therefore mandatory to use both matrix matched standards and suitable internal standards when analysing steel, metals and metal alloys.

Many high alloy steels have a matrix composition of Fe, Cr and Ni. If a nebulizer gas flow rate was kept lower than that giving a maximum analytical signal together with internal standardisation, a similar pattern of behaviour, regarding the matrix suppression effects, was obtained for the three elements. Thus trace elements in stainless steel could be determined by evaluation from only iron matched prepared standards.

On the whole ICP-MS is a powerful technique with a good accuracy for the determination of trace elements in metallurgical samples. Even phosphorus in low alloy steel can be determined at trace level concentrations.

REFERENCES

1. B. Berglund, R.W. Karlsson and Ch. Wichardt, <u>Fresenius Z. Anal. Chem.</u>, 1988, <u>330</u>, 498.
2. S.H. Tan and G. Horlick, <u>J. Anal. At. Spectrom.</u>, 1987, <u>2</u>, 745.
3. I. Gustavsson and H. Larsson, Technical Report from the Swedish Institute for Metals Research, Stockholm, IM-2794, 1992.
4. I. Gustavsson and H. Ekstroem, Technical Report from the Swedish Institute for Metals Research, Stockholm, IM-Report, in preparation.
5. A.R. Date and A.L. Gray, 'Inductively Coupled Plasma Mass Spectrometry', Blackie, Glasgow and London, 1989, Chapter 1, p. 31.

Determination of Antimony and Other Impurities in High Purity Copper by ICP-MS

I. A. Rigby and S. C. Stephen

ICI SPECIALTIES, SPECIALTIES RESEARCH CENTRE, BLACKLEY, MANCHESTER

INTRODUCTION

Trace metal impurities are detrimental to the physical and electrical properties of copper and limits for these are precisely specified[1]. Antimony is of particular interest and only a few papers on the use of ICP-MS for the determination of this and other impurities in copper appear to have been published. In these the impurities have been isolated by co-precipitation[2-4] or the copper has been removed from solution[5] prior to measurement by ICP-MS. In this work the direct introduction of copper solution into the ICP-MS has been studied together with flow injection and solvent extraction to remove the copper matrix from solution.

EXPERIMENTAL

Instrumentation. The instrument used was a commercially available inductively coupled plasma mass spectrometer, a VG Plasmaquad PQ2 Turbo Plus (VG Elemental Ltd., Winsford, Cheshire, UK). Operating parameters are given in Table 1.

Reagents. BDH Spectrosol 1000ppm stock solutions (ex Merck) were used to prepare all calibration solutions. Aristar nitric acid was used for sample preparation and de-ionised distilled water was used for all solutions. Indium was used as internal standard in all ICP-MS measurements. For the solvent extraction experiments a 5% m/v solution of 2-hydroxy-5-nonylbenzaldoxime in Escaid 100 (a standard diluent for solvent extraction of metals ex Essochem) was used.

Reference materials. Standard reference materials SRM 394 and SRM 395 ex National Institute of Standards and Technology (Gaithersburg, MD, USA) were used as reference samples.

Table 1.   Operating conditions for ICP-MS

Power	1350 W
Cool gas	13.5 l min^{-1}
Auxiliary gas	0.9 l min^{-1}
Nebuliser gas	0.9 l min^{-1}
Sample uptake	0.8 l cm^3 min^{-1}
Flow injection loop	500 µl
Mode	Pulse counting
Dwell time (scanning)	320 µsec
(peak jumping)	10240 µsec
Sweeps	100

Sample preparation. Samples were prepared by weighing 0.5g sample into 100 cm^3 beakers and adding 5 cm^3 water and 5 cm^3 nitric acid. The beakers were warmed on a hot plate until dissolution was complete and then allowed to cool. The solutions were transferred to 100 cm^3 graduated flasks and diluted to volume with water. Aliquots of 20 cm^3 were transferred to 50 cm^3 graduated flasks. These were spiked with indium solution to give a final indium concentration of 10 ppb and diluted to volume with water.

For the solvent extraction experiments 10 cm^3 aliquots of the sample solutions were transferred to a separating funnel and 10 cm^3 of the 2-hydroxy-5-nonylbenzaldoxime in Escaid 100 solution added. This was shaken for 30 seconds, allowed to separate and the solvent phase discarded.

RESULTS AND DISCUSSION

Limits of Determination. Initial experiments were carried out to assess the limits of determination which could be achieved. The instrument manufacturer recommends a maximum total solids content for sample solutions of 2g l^{-1}. Table 2 shows values for Limits of Determination calculated using 10 standard deviations on the blank and assuming a 2g l^{-1} sample solution. Data was acquired in Peak Jumping mode. These limits were considered adequate for samples likely to be encountered in this laboratory.

Table 2.   Limits of Determination (10 sigma values)

Fe	2.4	ppm
As	0.09	
Ag	0.11	
Sb	0.08	
Pb	0.12	
Bi	0.09	

<u>Direct Nebulisation of Copper Solution</u>. To study
the effect of copper on the instrument response for the
elements of interest a sample solution containing copper
at a concentration of 2 g l^{-1} was introduced into the
instrument and spectra collected at five minute
intervals. It was observed that there was a significant
reduction in response for all the isotopes measured over
a period of one hour. After this period the signal
intensity stabilised giving approximately 10 to 50% of
the original intensity depending on the particular
isotope. At this point the quadrupole could be retuned to
give between 50 and 60% of the original intensities.
Plots of change in intensity vs time for the measured
isotopes are shown in Figure 1. Plots of signal intensity
of each isotope ratioed to the indium internal standard
intensity are shown in Figure 2. These show some
improvement in stability for antimony, arsenic, iron and
silver while bismuth and lead do not.

The loss of sensitivity could be attributed to
blockage of the ICP-MS interface due to the relatively
high solids content of the sample solutions. Examination
of the interface after these experiments showed no
evidence in reduction of orifice diameter of either the
sample cone or the skimmer cone. The skimmer cone was,
however, heavily coated with what was presumed to be
copper oxide. A repeat of this experiment nebulising
1% v/v nitric acid solution between sample measurements
did not give any improvements to signal stability.

<u>Flow Injection Analysis</u>. Flow Injection can be used
in ICP-MS to reduce the problems encountered with
solutions containing high levels of dissolved solids[5]. The
above experiment was repeated using flow injection rather
than continuous nebulisation. Dilute nitric acid (1% v/v)
was used as the carrier stream and 500 µl injections of
copper sample solution were made at five minute
intervals. Data was collected for each sample injection
in scanning mode. Plots of intensity vs time for the
isotopes of interest are shown in Figure 3 and plots of
intensities ratioed to indium are shown in Figure 4. The
use of flow injection does give some improvement in
signal stability, with reasonably stable intensities
after the first ten minutes. The amount of contamination
collecting on the skimmer cone was also reduced.

To test this approach determinations were carried
out on the reference samples SRM 394 and SRM 395. The
mean values of duplicate measurements are shown in Table
3. The results for Sb, Pb and Bi are in good agreement
with certified values, however results for Fe, As and Ag
show poor agreement. It was noted that the elements not
in agreement with certified values show different drift
trends in intensity and intensity ratios compared to the
elements that show good agreement. Use of a second
internal standard to cover the lighter masses could
possibly give some improvement but this was not pursued.

Figure 1 **Direct nebulisation of copper solution**

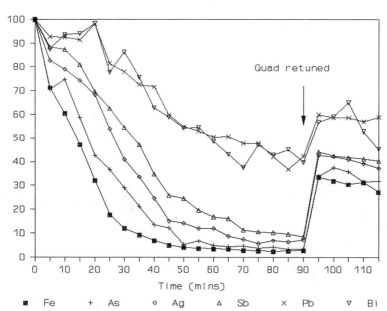

% Change in Intensity vs Time

Quad retuned

Time (mins)

■ Fe + As ◇ Ag △ Sb × Pb ▽ Bi

Figure 2

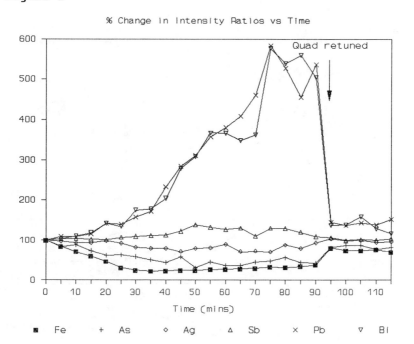

% Change in Intensity Ratios vs Time

Quad retuned

Time (mins)

■ Fe + As ◇ Ag △ Sb × Pb ▽ Bi

Figure 3   Flow injection of copper solution

FIA % Change in Intensity vs Time

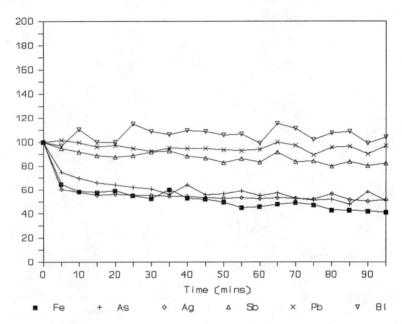

Figure 4

% Change in Intensity Ratios vs Time

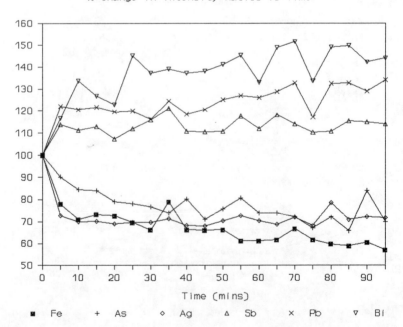

Table 3.   Results for SRM 394 and 395 by Flow Injection

	SRM 394		SRM 395	
	Found	Certified	Found	Certified
Fe	71	147	86	96
As	7.0	2.6	5.0	1.6
Ag	86	51	27	12.2
Sb	3.6	4.5	7.6	8.0
Pb	25	26.5	3.3	3.25
Bi	0.28	0.35	0.44	0.5

Solvent extraction. Solvent extraction has been widely used in sample preparation and was investigated for the removal of copper in the sample solutions. The Mining Chemicals Section of Specialties Research Centre have developed a ligand, 2-hydroxy-5-nonylbenzaldoxime, for the selective extraction of copper. It was shown that a 30 second extraction of 2 g l^{-1} copper solution using a 5% m/v solution of this ligand dissolved in Escaid 100 solvent could remove over 90% of the copper. Recovery experiments were carried out by preparation of solutions of known concentrations of the elements of interest followed by extraction of the solutions using a 5% m/v solution of 2-hydroxy-5-nonylbenzaldoxime in Escaid 100. The results of these experiments are shown in Table 4. For all the elements excellent recovery was obtained following extraction.

The solvent extraction procedure was applied to the reference sample solutions from SRM 395 and typical results are shown in Table 5. Good agreement was obtained between the concentrations found and the certified values for As, Ag, Sb and Bi. Agreement between the results found for Fe and Pb was poor and needs further investigation.

Table 4.   Recovery of elements following extraction

	Standard Concentration	Measured Concentration after extraction	Recovery %
Fe	200	212	106
As	4	3.9	97.5
Ag	80	83	104
Sb	10	10	100
Bi	2	1.9	95
Pb	40	38	95

Table 5.   Results for SRM 395 following extraction of
           copper

	Found	Certified
Fe	56	96
As	2.5	1.6
Ag	13	12.2
Sb	7.5	8.0
Pb	7.6	3.25
Bi	0.65	0.5

CONCLUSION

Studies of the direct introduction of high levels of
copper in solution have shown that severe interferences
can be produced. These have been attributed to interface
contamination arising from the copper matrix. Flow
injection partially alleviates these interferences but
does not give satisfactory results for all the elements
measured. Matrix removal using solvent extraction gives
good agreement with certified results on most elements
determined in the reference samples. Further work needs
to be carried out on the determination of Fe and Pb.

REFERENCES

1. ASTM B115-91 (02.01)

2. Y. Nakamura and T. Fukuda, Bunseki-Kagaku, Jan 1990,
   39(1), T17-T21

3. H. Umeda, I. Inamato and K. Chiba, Bunseki-Kagaku,
   Mar 1991, 40(3), 109-114

4. K. Chiba, I. Inamato and M. Saeki, J. Anal At.
   Spectrom., 1992, 3, 115

5. C.J.Park, S.R.Park, S.R.Yang, M.S.Han and K.W.Lea,
   J. Anal At. Spectrom., 1992, 4, 641

6. R.C.Hutton and A.N.Eaton, J. Anal At. Spectrom. 1988,
   4, 547

Lead Isotope Analysis of Nuragic Bronzes and Copper Ores by ICP-MS

E. Angelini and F. Rosalbino
DIP. SCI. MAT. ING. CHIM., POLITECNICO DI TORINO, ITALY

C. Atzeni and P. F. Virdis
DIP. ING. CHIM. MAT., UNIVERSITA' DI CAGLIARI, SARDINIA

P. Bianco
LABORATORIO CHIMICO CAMERA COMMERCIO, TORINO, ITALY

1 INTRODUCTION

The provenance studies of raw material of archaeological metal artifacts are extremely important in archaeometry, almost like the absolute dating of the object itself.

An example is related to the historical implications of the presence in Sardinia of the copper 'ox-hide ingots' (XV-X century b.C.): five ingots constituted by almost pure copper were discovered in the last century (in 1857) in the archeological site of Serra Ilixi near Nuragus (Nuoro), nowadays hundreds of ox-hide ingots have been found in the eastern Mediterranean Sea, moreover a mould was discovered in Siria.

The study of these metallic artifacts could clarify the metallurgical technique and the trades in the Mediterranean Sea from XV century b.C. (ox-hide ingots were painted in Egyptian tombs) to XII century b.C. (at least in the aegean-anatolic area).

Several hypotheses have been formulated on the history of the ox-hide ingots: they could be a metallurgical production of the Middle Bronze Age of an isle rich of mineral deposits or they could have arrived in Sardinia transported by a ship wrecked on its coasts (some of them show aegean-anatolic inscriptions and were found under the oldest megalitic towers) and could be successively hoarded up for several centuries by natives not able to utilize them until the Late Bronze Age.

The chemical determination of trace elements seems to be not very useful for the provenance studies on metallic artifacts deriving from complex high temperature treatments.

The recently developed lead isotope technique provides a more promising direct analytical method for linking artifacts with ore deposits.

Ages and mineral compositions of the various deposits sometimes are so different that they can be used as a fingerprint of the deposit. Of course the lead isotope ratio analysis not always gives a definitive answer, for example in the case of recasted objects.

Dealing with the Sardinian Prehistoric Age the lack of data and of sources is to be taken into account: several determinations have been carried out by Gales, who introduced this methodology in the mediterranean archeometallurgy[1,2].

The field of composition of Sardinian minerals has been defined on the basis of 91 samples of copper and lead ores; unfortunately this field partially overlaps with the cypriot field of copper minerals, mostly in the case of more recent ore deposits.
The few other determinations at disposal, performed by the researchers of the Max Planck Institute of Mainz, are in disagreement with some data obtained by Gales. For these reasons it is particularly interesting to compare the data coming from different laboratories.

In this study the lead isotope analysis technique has been applied to copper minerals of Sa Duchessa, Calabona, Funtana Raminosa and Montevecchio deposits, and to bronzes of the Museo Archeologico Nazionale of Cagliari, that are considered by archaelogists to be one of the most interesting witnesses of the Nuragic Age, that did not know writing and did not appreciate painting[3].

2   EXPERIMENTAL TECHNIQUE

The lead isotope analysis has been performed by means of plasma source mass spectrometry (ICP-MS VG Plasmaquad) on few mg taken from copper minerals, bronzes and lead objects. In order to reduce as much as possible the damaging of the bronzes, the sampling has been performed on the casting runners at the bottom of the statues.

The samples have been dissolved in nitric acid, and after a previous acquisition, diluted to 100 ppb Pb concentration; the acquisition has been carried out between 201 and 210 m/z; NIST SRM 981 (Common lead) has been used for calibration and mass bias correction. The results are average values between ten replicates.

The determinations have been performed on the four lead isotopes : ^{204}Pb, ^{206}Pb, ^{207}Pb, ^{208}Pb. As generally accepted ^{204}Pb comes totally from primordial lead, while ^{206}Pb, ^{207}Pb, ^{208}Pb partially come from the radioactive decay of ^{238}U, ^{235}U, ^{232}Th. ^{202}Hg was monitored and used for the calculation and correction of ^{204}Hg contribution to ^{204}Pb isotope abundance in Hg containing samples.

3   RESULTS AND DISCUSSION

In Fig.1 are shown the minerals examined and the ore deposits of provenance.

The results of the lead isotope analysis carried out on the above cited minerals are shown in Fig.2: they fall in the Sardinian mineral field, defined on the basis of Gales' data, although some specimens widen the boundaries.
These results could be useful in evaluating the dispersion of data referred to minerals of the same deposit.

In Fig.3 lead isotope ratios obtained by different Authors on minerals coming from Sa Duchessa, Calabona, Funtana Raminosa and Montevecchio, are shown, the dispersion is noticeable; on the other hand previous data obtained by geologists[4] showed that Montevecchio ore deposits have an isotopic age different from the geological one.

The lead isotope analysis of ox-hide and bun ingots are shown in Fig.4. Most data come from Gales' analyses (22 samples) and fall close to the Cypriot field, but as said above, they could belong to the Sardinian field too.

Our data partially agrees with Gales' data, while are in total disagreement with Swainback's data, in spite of the same provenance site of the samples.

The examination of data obtained leads to the hyphothesis that ox-hide and bun ingots belong to two different groups (different deposits and different ages): the former being constituted by highly pure copper cannot derive from the refining of the latter one. Both of them could have been produced with Sardinian minerals, but for the ox-hide ingots the Cypriot origin cannot be excluded, Fig.4.

The bronzes of the Museo Archeologico Nazionale of Cagliari submitted to the lead isotope analysis are listed in Fig.5.

The results, shown in Fig.6, indicate that the

MINERALS	DEPOSITS
1 GALENA,	Montevecchio
2 GALENA,	Santadi
3 CHALCOPYRITE,	Funtana Raminosa
4 CUPRITE,	Calabona
5 COVELLITE,	Calabona
6 CHALCOPYRITE,	Sa Duchessa
7 MALACHITE,	Rosas
8 CUPRITE,	Montevecchio
9 CHALCOPYRITE,	Tertenia
10 MALACHITE,	Ozieri

Figure 1 Sardinian minerals examined with indication of the ore deposit on the Sardinia map

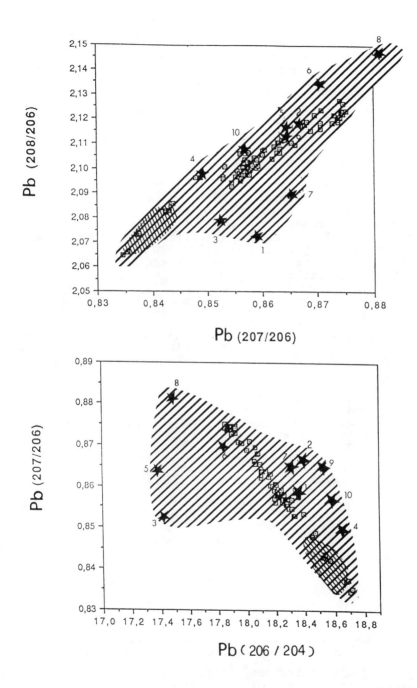

<u>Figure</u> 2 Lead isotope analysis of minerals of Figure
1(★); Gales' data (▢); Sardinian mineral field
▨ , Cypriot mineral field ▨ .

statues belong to a unique context and fall in the
Sardinian field, closer to the bun-ingots field with
respect to the ox-hide ingots field. The values
obtained by Gales on bronzes coming from Santa Maria
in Paulis are rather similar.

     In the opinion of many archeolgists the bronzes
could be attributed to the first Iron Age (IX-VII
centuries b.C.); this hypothesis seems to be confirmed by the observation that for their production
copper coming from ox-hide ingots has not been employed.

     The results of lead isotope analysis of some
lead artifacts (sheets, little ingots, cramps),
Fig.6, allow us to state that these objects also belong
to an homogeneous group, notwithstanding their attribution to different ages. The artifacts coming from
Antigori probably belong to 1300-1200 b.C., while the
objects coming from Santa Barbara are attributed to
IX-VII centuries b.C., Iron Age.

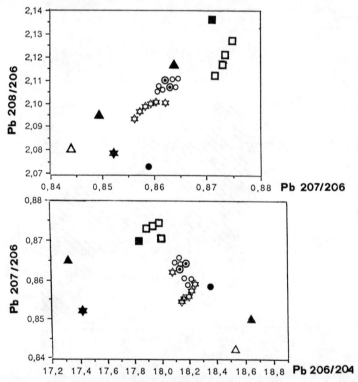

Figure 3   Lead isotope analysis of different researchers on minerals of the same deposit. Montevecchio: our data(●), Gales(O), Swainback(⊙); Funtana
Raminosa:our data(✹),Gales(✩); Calabona: our data(▲),
Gales(Δ); Sa Duchessa : our data(■), Gales(□)

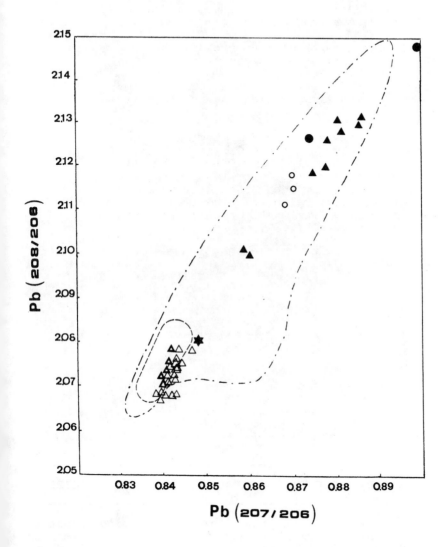

Figure 4 Lead isotope analysis on ox-hide and bun
ingots obtained by different Authors.
Ox-hide ingots: Gales, Crete (▲), Cyprus (▲),
Sardinia, Villanovaforru (△); our data, Sardinia
Villanovaforru (★).
Bun ingots: Gales, Villanovaforru (O); our data,
Villanovaforru (●).

BRONZES OF THE ARCHAEOLOGIC MUSEUM OF CAGLIARI

```
 1 - n. 84 : Praying soldier  with hanging shield
 2 - n.110 : Devil
 3 - n. 93 : Praying soldier with hanging shield
 4 - n.104 : Hero with four eyes and four arms
 5 - n.121 : Praying woman
 6 - n.128 : Soldier with hanging shield
 7 - n.116 : Praying woman with long plaits
 8 - n.207 : The yoke
 9 - n.216 : Ox standing
10 - n.144 : Offering woman with short cloak
11 - n.145 : Priest with stole
```

Figure 5 Bronze n.104 and bronzes examined

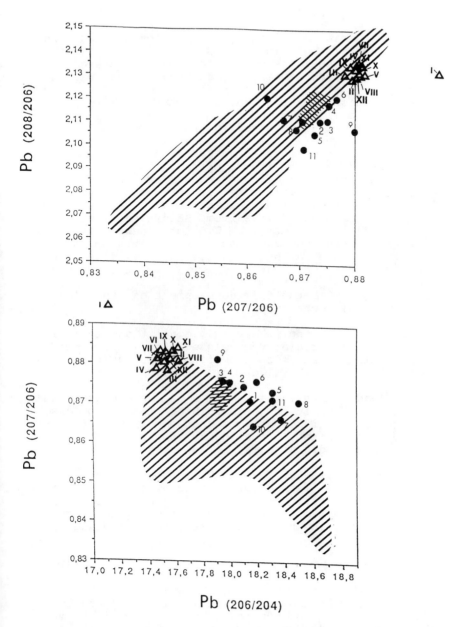

Figure 6 Lead isotope analysis of bronzes of Fig.5 (●), samples 1-10, and of lead artifacts (▲) from Santa Barbara, Bauladu (samples I-III) from Sarroch, Nuraghe Antigori (samples IV-VII), from Villanovaforru (samples VIII-XII);
Gales' data on Antigori lead objects ⊠⊠ and Sardinian mineral field ▨ .

4   CONCLUDING REMARKS

- The Sardinian field of isotope ratios of lead and copper containing minerals is defined by nearly one hundred analyses; its wide extension is related to the complex minerogenesis of the island. The partial overlapping with the Cypriot field prevents an easy discrimination. Moreover the literature data are not referred to minerals found in ore deposits, whose activity is well-documented in prehistory.

-    The available isotopic analyses on ingots show that ox-hide and bun ingots probably come from different deposits and different ages: ox-hide ingots, constituted by highly pure copper, cannot derive from the refining of bun-ingots.
Both of them could derive from Sardinian minerals, and ox-hide ingots also from Cypriot minerals. Fragments of ox-hide and bun ingots are sometimes associated in the same archeological site thus accrediting the hypothesis of an hoarding of the ox-hide ingots and of their employment in a later age (Late Bronze Age) with the full development of the local metallurgical techniques.

- The statues belong in a unique context and fall in the Sardinian field, very close to the bun ingots field; this finding should confirm their attribution to the Iron Age (IX-VII centuries B.C.).

-    The lead artifacts belong in a unique context and fall substantially in the Sardinian field; because they belong to different ages and come from different archeological sites, a continuity in the exploitation of the mineral deposits may be supposed.

-    The application of the lead isotope analysis is an interesting method for characterizing metallic archeological artifacts, however it cannot be considered conclusive for provenance studies, and should be integrated by trace analysis and archeological elements.

REFERENCES

1. N.H.Gale, Review of the application of lead isotope analyses to provenance studies, Proc. Symp. Archaeometry, Ed. Maniatis. Athens 1986.
2. N.H.Gale, Z.A. Stos-Gale, "Studies in Sardinian Archaeology", III, B.A.R., Ed. Balmuth, 1987.
3. G.Lilliu, "Le sculture della Sardegna Nuragica", Ed. La Zattera, Cagliari, 1966.
4. P.Zuffardi, Symposium A.M.S., Cagliari Iglesias, 1965, Ext.Abs.Sez I, A-2, 39
5. I G.Swainback, T.J.Shepherd, R.Caboi, R.Massoli-Novelli Period.di mineralogia-Roma, 51, 1982, 275

Measurement of Isotope Ratios Using ICP-MS

P. J. Turner
FINNIGAN MAT LTD., UNIT 7, ASHER COURT, LYNCASTLE WAY,
APPLETON, WARRINGTON, CHESHIRE WA4 4ST, UK

Introduction.

Inductively coupled mass spectrometer systems are increasingly being used for isotope ratio measurement in various areas of elemental analysis[1]. Such measurements have previously been made by thermal ionisation mass spectrometry, a technique which, although capable of providing extremely precise and accurate results, requires that samples for measurement undergo a difficult and time consuming preparation process. I.C.P. mass spectrometry with its minimal requirements for sample preparation is an attractive alternative provided that due consideration is given to the intrinsic limitations of the technique.

Measurement Considerations.

There are various aspects which must be considered to determine whether the I.C.P. technique is suitable for the measurement of isotope ratios for a given application. These are:
1. Sample quantity available.

It will be shown later that the overall efficiency of the I.C.P. is some 2 to 3 orders of magnitude lower than that of the thermal ionisation mass spectrometer. Where sample quantity is small, the I.C.P mass spectrometer may be unable to produce a useful result.

2. Required accuracy and precision of the result.

Different applications require different accuracies for the result. Typical requirements for geological dating purposes are shown in table 1.
It can be seen that there are two types of measurement requiring very different levels of measurement precision. For Sr and Nd there are isotope pairs that are regarded as invariant in nature and which may therefore be used to correct for measurement bias.

Element	Isotopes	Normalised	Accuracy	Ratio
Pb	204,206,207,208	n.a.	.1%	All
Sr	84,86,87,88	86/88 = .1194	.001%	87/86
Nd	142,143,144,145,146	146/144 = .7219	.0005%	143/144
Os	186,187,188,192	188/192 = .3261	.5%	187/186
Th	230,232	n.a.	.1%	230/232

Table 1.   Geological Dating Applications.

Reproducibilities of better than 10 ppm may be obtained using thermal ionisation techniques in conjunction with magnetic sector analysers. This degree of accuracy is generally regarded as necessary for these isotope dating measurements.For lead osmium and thorium the experimental requirement is somewhat less stringent, being typically of the order of 0.1%.

For other areas of application, typically involving isotope dilution measurements, precisions and accuracies of the order of 0.1% to 0.5% are generally all that is required. Such measurements might be used for the following areas of research.

a) Metabolic studies involving elements such as Fe,Ni,Zn,Cu which have two or more stable isotopes which can be used for isotope dilution studies. By suitably spiking with an aliquot of the element which has been enriched in the abundance of one of its isotopes and measuring the isotope ratio of the resulting mixture, the concentration of that element may be calculated.These concentration measurements can then be used to derive information relating to metabolic time response and elemental distribution.

b) Environmental studies. Again, the method of experimentation is to spike samples with aliquots of elements containing non natural isotopic abundances and to measure the isotope ratios of the resulting mixtures. Concentrations and distributions of elements in the environment can be obtained in this manner.

c) General isotope dilution analyses for chemical quantification.The diagram below shows the principle of this method of analysis. The sample in this case is assumed to have two isotopes A and B having abundances which are represented by the upper line. The spike is assumed to have an isotopic composition represented by the middle line and the result of mixing the two components is shown by the lower line. It can be shown that the concentration of the element in the original sample can be calculated simply by measuring the isotope ratio of the mixture, assuming that the ratios for the sample and the spike are known, together with the volumes of the spike and sample solutions and the concentration of the spike solution. The great advantage of this method is that the calculation depends only on the measurement of an isotope ratio and no knowledge is necessary of the recovery of the sample after initial spiking.

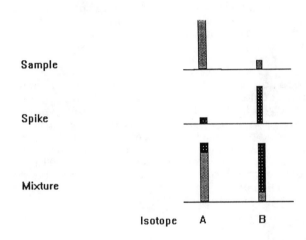

Sample

Spike

Mixture

Isotope   A   B

$$\text{Conc.sample} = \frac{\text{Vol.spike} \times \text{Conc.spike} \times \text{At.wt.sample}}{\text{Vol.sample} \times \text{At.wt.spike}}$$

$$\times \frac{\left(1 - r_{mix}\!\left(\frac{A}{B}\right) \times r_{spike}\left(\frac{B}{A}\right) \right) \times \text{Ab.spike A}}{\left(r_{mix}\!\left(\frac{A}{B}\right) \times r_{sample}\left(\frac{B}{A}\right) - 1 \right) \times \text{Ab.sample A}}$$

Isotope Dilution Calculation.

3. Basic limitations due to the use of a quadrupole mass filter.

Quantification of the intensity of the peak of a given mass as presented by the quadrupole presents some difficulties due to the rounded nature of the peak. It is necessary to know the absolute accuracy of a given isotope ratio measurement as well as having an adequate precision on this measurement.

4. Matrix effects can severely distort the isotope ratios obtained. These effects can be of several types.

a) Isobaric interferences need to be considered to make sure that the presence of an isotope of a different element does not distort the measured ratio.

b) More difficult to detect is the possible presence of molecular interferences under the isotope peaks to be measured. The intensity and mass of these molecular ions depends on the specific elements present in the samples to be measured.

c) Change of slope of mass bias curve due to presence of high concentrations of elements in the sample solution. The energy distribution of ions originating in the plasma can be changed significantly in the presence of for example large amounts of Na in the sample solution. If there are energy filtering elements in the ion transfer optical system, the relative transmission efficiencies of ions of different masses will change since the energy of the ions in the system is approximately proportional to their mass.

5. Magnitude of ratio to be measured.

Problems can arise if a very large ratio is to be measured. In order to have sufficient beam intensity available to obtain a precise measurement of a minor isotope, the intensity of the major isotope may have increased to a level where it can no longer be measured on the electron multiplier. Under these circumstances it can be useful to have a Faraday detector which can be used to measure the high intensity ion beams without running into problems of saturation. With a dual collector system a useful strategy is to spike the sample with an appropriate concentration of a second element so that a peak of an intermediate intensity is available which can be measured on both the multiplier and the Faraday collectors.

System Efficiency.

The sensitivity of an I.C.P. mass spectrometer is of the order of 10^8 ions per second per ppm weight for a mid mass range element using a standard concentric nebuliser. Sample consumption rate is 1 ml per minute, so that the overall efficiency of the analyser system is 1 part in 10^6. With an ultra-sonic nebuliser this can be improved by a further order of magnitude to 1 part in 10^5. This should be compared with an efficiency of .1% to 10% which is typically obtained for thermal ionisation mass spectrometers. The relevant parameters for the two techniques are shown in table 2.

Ionisation Method	Efficiency	Sample size	Precision
Thermal	10^{-1} to 10^{-3}	10^{-8} gm	.0005%
ICP	10^{-5}	10^{-6} gm	.1%

Colletor type: Faraday.

Table 2.   Measurement Efficiencies.

For samples such as uranium, figures comparable to thermal ionisation may be obtained from I.C.P systems. There is no shortage of sample, and only modest precisions are required for the ratios. For geological samples such as Nd, sample sizes may only be a few nanograms, and the precision required for the measurements is 2 to 3 orders of magnitude better than that currently obtainable from quadrupole I.C.P mass spectrometers. Multiple collection using a double-focusing sector mass spectrometer will improve precision, but acceptable quality ratios will not be obtained from small samples until overall efficiency can be improved substantially.

<u>Measurement of Ag Isotope Ratios.</u>

The spectrum from a 2 ppm Ag solution is shown below. The individual isotopes are well resolved from each other and the top of the peaks are rounded as is typical from a quadrupole mass analyser. Peak intensities are measured by adding together the individual counts in each of the channels across the centre portion of the peak. The number of channels used is user specifiable and depends upon what compromise is considered to be optimum. Since the ion beam from the I.C.P. is intrinsically slightly unstable, it is desirable to move as rapidly as possible from one peak to another to average out the effects of beam fluctuations.On the other hand, in order to get the best measure of the beam intensity a large number of measurement points across the top of the peak is necessary, resulting in a longer cycle time between

peaks and a consequent degradation in the precision of
the measurement due to beam instability.
Typical results from silver are shown in table 3.

Repeat No	Ratio
1	1.0776
2	1.0821
3	1.0783
4	1.0782
5	1.0799

$\overline{X}$ = 1.0792   Collector : Faraday

s.d = .17 %   Dwell : 8 msec per channel

s.e. = .08 %   Channels per amu : 16

                                           Passes : 140

Table 3.

The standard deviation between each measurement is .17%
and the standard error on the combined ratios is .08%.
In this particular case the observed value is close to
the "true" value of 1.078. If the same measurement is
made on a series of different instruments, it is found
that whilst individual systems give answers which are
precise to better than .1%,the scatter of ratios
between instruments is somewhat larger, see table 4.

System No	Ag isot ratio	Precision
1	1.0792	.08%
2	1.0697	.15%
3	1.0706	.08%
4	1.0600	.09%
5	1.0628	.04%
6	1.0678	.07%

$\overline{X}$ = 1.0684

s.d. = 0.6%

Table 4.
    Inter-system reproducibility of Ag 107/109 ratio.

If isotope ratios are measured on other elements across
the periodic table, the uncorrected values are found to
be typically within 1-2% of the true values,
irrespective of the mass or the mass difference of the
isotopes. Values for Li,Mg,and U are shown in table 5.

Element	Ratio	Mean	R.S.D.	True
Mg	26/24	.1462	.09%	.1396
Mg	25/24	.1285	.16%	.1266
U	235/238	.9982	.62%	.9998
Li	6/7	.0786	.19%	.0807

Table 5.   Isotope ratios as function of mass.

The overall system bias for the SOLA is clearly small
as a function of mass, but there is a small systematic
bias for each ratio as measured. Under normal
circumstances this bias is corrected for by means of
standards, but however there may arise situations where
it is not possible to run standards. In the case of
isotope dilution it would be useful to know that
reported ratios are within .2% of the true ratios
without the necessity for repeated running of
standards.

Sources of error.

Possible sources of error for the measurement of
isotope ratios are listed below. These fall into two
categories:

Systematic.
1.Interface and transfer lens effects.

2.Quadrupole measurement protocol.

3.Sample introduction system.

Random.
1.Beam instability.

2.Detector noise.

Transfer lens effects.

The interface and transfer lenses must be designed so that bias effects due to space charge and ion energy distribution are minimised. The SOLA uses a high voltage ion optical transfer system which keeps these effects to as low a level as possible. The energy distribution of the ions travelling through the interface is such that the mean energy of a given ion is approximately proportional to its mass - this is a consequence of the ions travelling through the interface with nearly constant velocity. If the transfer window is insufficient, then effects such as that shown below will occur.

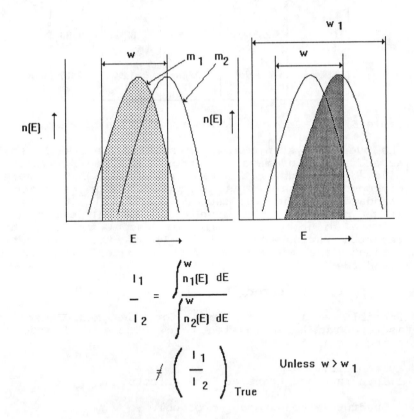

$$\frac{I_1}{I_2} = \frac{\int^w n_1[E]\ dE}{\int^w n_2[E]\ dE}$$

$$\neq \left(\frac{I_1}{I_2}\right)_{True} \qquad \text{Unless } w > w_1$$

Figure 1.   Energy transmission effects.

Here two ions of masses m_1 and m_2 have been considered. These have energy distributions $n_1(E)$ and $n_2(E)$ which are displaced with respect to each other depending on mass. If the transmission window is width w as shown then only a part of the total number of ions of each mass will be transmitted through to the analyser. The relative proportion of each ion beam will depend on the

position of the energy window as will the measured
value for the ratio of the intensities of the beams.
The true ratio of the intensities of the beams will
only be obtained if $w > w_1$. Systems incorporating
strongly filtering optics e.g. a Bessel Box filter will
be susceptible to this effect. Moreover, since energy
distributions depend on the matrix content of the
sample in the plasma, different isotope ratios will be
obtained from different matrices.
Since ion transmission is not a strong function of mass
in the SOLA, isotope ratio bias due to this effect is
small.

Measurement Protocol.

The peaks produced by a magnetic mass spectrometer
system and a quadrupole mass spectrometer system are
shown in figure 2.

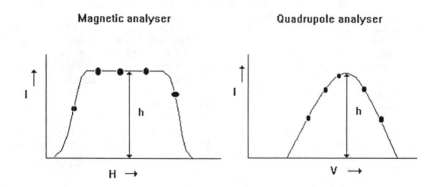

Figure 2.
  Peak shapes for magnetic and quadrupole analysers.

Magnetic systems are set up to produce flat topped
peaks as the the magnetic field is scanned over the
various masses. The intensity of the peak is then
simply the height h of the peak, which is independent
of the position at which the measurement is made. This
is important,since the digital scan used to drive the
value of the magnetic field will never actually sample
the peak at its exact centre.

A severe problem arises in the case of the quadrupole
peak since the top is rounded. At no point is there a
region where the intensity is independent of the scan
position, and since the digital scan will only produce
intensities from discrete positions over the peak, no
measurement will be made from the true top of the peak.
Various methods may then be used to quantify the
intensity of the peak. These are:

1) Peak height.

2) Peak area.

3) Curve fitting.

The choice of method is then a compromise between accuracy and speed of measurement. If a best estimate of the peak height is to be obtained, then a fine digital scan must be carried out over the top of the peak. This will reduce the error between the measured maximum and the true maximum of the peak. However this will inevitably mean that more time is spent in the measurement of a given peak, and the cycle time through a sequence of peaks will be increased. The effect of time instabilities of the ion beam will then reduce the precision of the measurement. Using an area calculation does not help the problem of finding the true maximum of the peak. Whilst the signal to noise ratio may be improved, there still remains the possibility of retaining a bias in the measurement. If only one point is used per peak, then the measurement protocol reduces to the so-called peak hopping method which is intrinsically very rapid but is obviously susceptible to errors introduced by the fact that it will be impossible to hit the exact tops of the peaks. A further important disadvantage of this method is that a different bias will be introduced for different isotope ratios since the point at which a measurement is taken on one peak will be at a slightly different relative position on the other peaks. It will be difficult to apply the usual correction procedure which is used in magnetic analysers whereby a constant mass bias per amu is assumed.

Curve-fitting appears to offer a partial solution to this problem. A schematic peak from a quadrupole analyser is shown in figure 3.

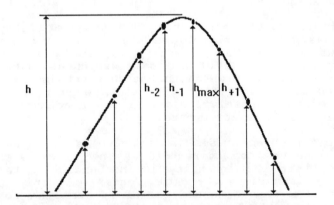

Figure 3.   Digital scan on quadrupole peak.

A typical number of points per a.m.u. for a scan is 16, and this would result in intensities being measured at each of the points as shown. In general there will be one maximum, not the true peak maximum, and two somewhat lower measurements taken on either side of this maximum. If it is assumed that peak may be approximated by a second order curve over the region near its maximum, and the points at which it is sampled, then the true maximum is given by the following expression:

$$ h = \frac{\left(h_{-1} - h_{+1} \right)^2}{8 \left(2h_{max} h_{-1} - h_{+1} \right)} + h_{max} $$

The result is independent of the exact position of the digital scan across the peak. The expression may also be used to estimate the error to be expected from a single point measurement. If h_{-1} and h_{+1} are assumed to be .95 and .85 then the correction term is 0.6% which is of the same order as the errors reported for the Ag isotope ratio results.

Conclusions.

1. Using a quadrupole mass spectrometer isotope ratio measurements using an I.C.P. ion source are accurate to 1 - 2 % without correction.

2. Precisions are typically .05%, provided beam size is not a significant factor.

3. Typical sample concentrations for isotope ratio measurements are 1ppm for the Faraday collector and 10 ppb for the multiplier collector.

4. The digital scan on the quadrupole peak is a significant source of error since in general the true peak maximum will never be measured.

5. It should be possible to reduce errors using a simple second order curve fitting procedure over the top of the peak.

References.

1. D.W.Koppenaal, Analytical Chemistry, Vol 64, No12,June 1992,p 325.

Direct Determination of Uranium Isotopic Abundance by Laser Ablation ICP-MS

X. Machuron-Mandard
COMMISSARIAT A L'ENERGIE ATOMIQUE, INSTN, SECTION DE CHIMIE ANALYTIQUE, 91191 GIF-SUR-YVETTE, FRANCE

J. C. Birolleau
COMMISSARIAT A L'ENERGIE ATOMIQUE, BP 12, 91680 BRUYERES-LE-CHATEL, FRANCE

Abstract

Isotopic abundance of elements for inorganic compounds can be determined by applying the well-known Thermal Ionization Mass Spectrometry technique (TIMS). However, this method suffers time-consuming sample preparation and accurate calibration using expensive standard materials. Moreover, the amount of time required to record the data does not always satisfy the analytical chemist when only rapid and not very accurate results are needed. From this point of view, Inductively Coupled Plasma Mass Spectrometry (ICP-MS) can be considered as an efficient tool applied to rapid isotopic as well as multi-element analysis.

This paper deals with the isotopic analysis of uranium compounds deposited onto cellulosic filters as solid samples. Direct laser ablation sampling was achieved and gave rise to a local and microscopical isotopic abundance analysis. The procedure adopted and described in this paper implies neither calibration nor bias correction. The direct measurement of the area corresponding to the peak of each uranium isotope yielded good precision and accuracy (1 to 10 %) over a wide range of uranium 235 enrichment (0.7 to 93 atom %). Results obtained with different NBS standard materials are presented and illustrate the great potential of ICP-MS as rapid micro-analytical method.

Introduction

The determination of the isotopic abundance of elements can be achieved by thermal ionization mass spectrometry (TIMS). Very precise and accurate results are usually obtained by applying this technique to most inorganic or organo-metallic compounds encountered in the course of chemical research. However, this method suffers time-consuming sample preparation and implies accurate calibration of the spectrometer in order to obtain precise data and reliable results. Even if micro-sampling techniques have been developed for TIMS analysis,[1] numerous applications of this technique mainly concern the determination of average isotopic abundance because of macro-sampling achievement. In which case, the amount of analyte sampled is too great if compared to the size of isotopically heterogeneous particles. Isotopic abundance information about microscopic particles cannot generally be obtained by applying standard sampling procedures. Thus, due to the low spatial resolution sampling technique, mixed powders made of different isotopic grade materials can be regarded by TIMS as homogeneous materials having average isotopic abundance. However, it may be very important to obtain local isotopic information and to be able to discern microscopic particles with a different isotopic composition. In the nuclear field, this distinction has sometimes to be made between depleted, natural and enriched uranium oxide powders. Samples of "natural average isotopic abundance" could be synthetized by mixing together depleted and highly enriched uranium oxide powders. In which case, the knowledge of the isotopic abundance for individual particles is fundamental in order to account for the procedure adopted to produce the material analysed.

This paper deals with the microscopic determination of uranium isotopic abundance in order to make a clear and rapid distinction between close particles made of depleted, natural or highly enriched uranium. The aim of this work is to demonstrate that ICP-MS can be applied to solve such analytical problems in a very simple and efficient way. The analytical procedure adopted is briefly described and the results presented illustrate the ability of Laser Ablation Inductively Coupled Plasma Mass Spectrometry (LA-ICP-MS) to carry out such an analysis.

Experimental

Instrumentation

The work described herein was performed on a VG-Elemental PlasmaQuad II Turbo Plus ICP-MS modified for containment in a glove box and connected to a VG LaserLab Plus laser ablation sampling device. Infrared laser beam (wavelength 1064nm) was generated by a Nd:YAG optical road and focused onto the uranium oxide particles through the quartz window of a sample cell. This cell was flushed by approximately 1 L/min of argon carrier gas and the aerosol produced by laser ablation was injected into the plasma after passing through Nylon and Tygon tubing.

Reference materials

All uranium oxide isotopic standard materials were supplied by the National Institute of Standards and Technology (NIST, USA), except natural uranium which was provided by the Commissariat à l'Energie Atomique (CEA, France).

The NBS 612 glass standard material was used to tune the spectrometer and obtain maximum sensitivity for lanthanum (^{139}La) prior to the isotopic analyses.

The filter paper used as support to maintain the samples in the sampling chamber was made of cellulose. The uranium background signal was measured by applying the analytical procedure to clean filters. No significant uranium signal was detected under the standard experimental conditions adopted for uranium samples analysis.

Experimental procedure

The apparatus was kept in stand-by position for over 30 minutes to reach a stable thermal state prior to analysis. Then, tuning of torch position and lenses was performed by using NBS 612 standard material to which laser ablation sampling was applied. Lanthanum 139 was selected to tune the apparatus and gain maximum sensitivity. The operating parameters are presented in table 1.

The uranium oxide sample was deposited onto filter paper as minute and widely dispersed particles.

Laser ablation was achieved after selection of individual particles (1 to 100 μm diameter). Mass scanning spectra were recorded from 233 to 240 u (Table 1). Isotopic ratios were determined by direct measurement of peak area for each uranium isotope (Figure 1 and 2). No blank subtraction was required for calculations due to an intense uranium signal to background signal ratio. Two calculation procedures were applied in order to obtain the best analytical results. The first one consisted in direct peak area measuring and no bias correction for uranium isotopic abundance calculation (Table 2). The second procedure was based on bias corrections. The bias factors were determined for isotopes 234, 235 and 238 using natural uranium. Five replicate measurements were taken to improve the precision of bias factors. The results obtained by applying this second procedure are presented in table 3.

Results and discussion

The results presented in table 2 were obtained on the assumption that no difference in sensitivity exists between uranium isotopes. In fact, due to the very narrow mass range explored: from uranium 234 to uranium 238 and the very similar ionization energies for these isotopes; the spectrometer atomic signal response can be considered as equal for each uranium isotope. In that case, owing to the flat response curve of the spectrometer and as P.J. Turner already pointed out,[2] no bias correction is necessary - the mass peak area and the atomic amount of uranium are proportional. Thus, the normalization of area measurement leads in a very simple way to the determination of the isotopic abundance for each isotope. From the data presented in table 2, we can see that the precision obtained by applying this procedure is quite satisfactory. Major isotopes are determined with good precision (1 to 10% - difference between certified and obtained value over certified value). Obviously, due to low counting rates, minor isotopes give less precise results (1 to 20% for relative precision).

Table 1:   Operating parameters

Argon gas flows (L/min)	Coolant: 13 Auxiliary: 0.75
	Laser Sampling Chamber: 1
Plasma rf power (W)	Incident: 1250 Reflected: < 5
Ion optic settings	Extractor: 5.0 L1: 8.12 L3: 6.36
(potentiometer turns)	Collector: 6.8 L2: 4.56 L4: 2.82
	Pole Bias: 4.92
Detector type	Philips; Channeltron
Collector type	Pulse
Mass range	233 to 240 u
Acquisition time	20 s
Number of channels	512
Number of scan sweeps	2000
Dwell time	20 μs
Detector voltage	-2.5 kV
Laser device	Pulsed Nd:YAG (1064 nm);
	Spectron Laser Systems; SL400
Energy	105 mJ free-running (focused)
Luminous flux density	5×10^5 W/mm^2
Laser pulse frequency	1.5 Hz

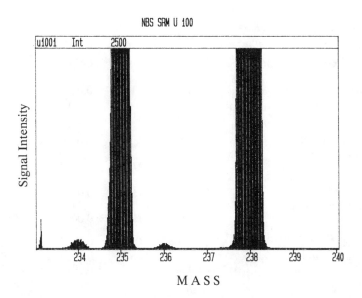

Figure 1:   Mass spectrum showing uranium isotopes and recorded after laser ablation sampling. (uranium oxide NBS U100).

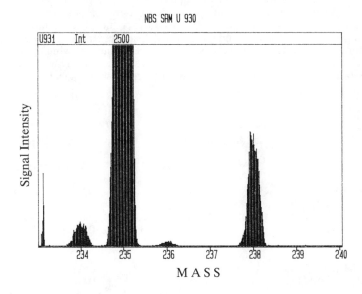

Figure 2:   Mass spectrum showing uranium isotopes and recorded after laser ablation sampling. (uranium oxide NBS U930).

Table 2:   Direct Isotopic Abundance Determination.

STANDARD MATERIAL	URANIUM ISOTOPES			
	^{234}U	^{235}U	^{236}U	^{238}U
Natural uranium (CEA)	0.006	0.63	0.001	99.4
Certified abundance [	0.0054	0.7202	< 0.0001	99.275
	± 0.0001	± 0.0002		
NBS SRM U100	0.084	10.9	0.039	89.0
Certified abundance [	0.0676	10.190	0.0379	89.704
	± 0.0002	± 0.010	± 0.0001	± 0.010
NBS CRM U500	0.58	50.9	0.092	48.4
Certified abundance [	0.5181	49.696	0.0755	49.711
	± 0.0008	± 0.050	± 0.0003	± 0.050
NBS SRM U930	1.21	93.1	0.22	5.5
Certified abundance [	1.0812	93.336	0.2027	5.380
	± 0.0020	± 0.010	± 0.0006	± 0.005

The previous data can be improved if bias correction is achieved as shown in table 3. Using natural uranium to determine bias factors, most of the results are more precise. However, no improvement is gained by applying bias correction procedure to very low level isotopes. This is probably due to very low ion signal intensity as well as poor accuracy for the bias factors.

The accuracy of the results was determined by applying Student-Fisher statistics to ten replicate isotope ratio measurements.[3] An estimate of the relative uncertainty calculated for 95% confidence limit is presented in figure 3 versus atomic isotopic abundance level. Of course, there is no doubt that TIMS or even multiple collector ICP-MS, development of which is currently under way,[4] are more suitable for obtaining very accurate results. However, the data presented in figure 3 are quite satisfactory with respect to the aim of this study.

Table 3:   Isotopic Abundance Determination with Bias Correction.

STANDARD MATERIAL	URANIUM ISOTOPES			
	^{234}U	^{235}U	^{236}U	^{238}U
Natural uranium (CEA)	0.020	0.82	0.022	99.1
Certified abundance [	0.0054	0.7202	< 0.0001	99.275
	± 0.0001	± 0.0002		
NBS SRM U100	0.076	11.1	0.037	88.8
Certified abundance [	0.0676	10.190	0.0379	89.704
	± 0.0002	± 0.010	± 0.0001	± 0.010
NBS SRM U930	1.03	93.3	0.20	5.4
Certified abundance [	1.0812	93.336	0.2027	5.380
	± 0.0020	± 0.010	± 0.0006	± 0.005

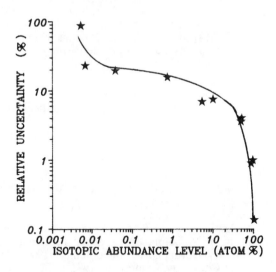

Figure 3:   Estimate of measurement uncertainty versus isotopic abundance
level for 95% confidence limit.

In order to estimate the detection limit of the technique described in the previous procedure, acidic aqueous solutions containing different amounts of uranium nitrate were deposited onto filter papers. The size of the deposit was estimated and the uranium surface concentration was calculated assuming that the absorption of the solution was homogeneous. The size of the hole produced by laser ablation was measured with an optical microscope (40 μm diameter). The minimum amount of natural uranium necessary to determine the isotopic abundance of isotopes 235 and 238 was estimated at 100 ng. This amount permits measurement of a significant signal for both isotopes and is necessary to obtain reliable results. This "detection limit" illustrates the great sensitivity of LA-ICP-MS.

Finally, it must be pointed out that laser ablation can give rise to local pollution due to dust production and redeposits. Under our experimental conditions, the distance between two particles made of different isotopic grade materials must be at least 3 mm to avoid any pollution of   microscopic samples.

Conclusion

The results presented in this paper demonstrate very strikingly the great potential of LA-ICP-MS as a rapid micro-analytical tool. The procedure adopted and described for isotopic abundance determination of microscopic particles illustrates the very simple way important information about nuclear materials can be obtained.

In addition to numerous applications of LA-ICP-MS reported in recent publications, new developments of laser ablation have been presented recently and give rise to very small crater size (typically 5 μm).[5] This progress tends to indicate that in the near future, improvements of this very efficient sampling method and the development of new ICP-MS machines (e.g. multiple collector ICP-MS) should offer the analytical chemist powerful tools for solving numerous problems that become increasingly difficult.

Acknowledgment

The authors thank Mr. V. COSTA for his critical reading of the manuscript.

References

1. T. Hirata, Proc. Third Int. Conf. Plasma Source Mass Spectrometry, Durham, 13-18th September 1992.

2. P.J. Turner, 'Applications of Plasma Source Mass Spectrometry', Eds G. Holland and A. N. Eaton, Proc. Second Int. Conf. Plasma Source Mass Spectrometry, Durham, 24-28th September 1990, Published by the Royal Society of Chemistry, Cambridge, 1991, p. 71.

3. G.H. Jeffery, J. Bassett, J. Mendham and R.C. Denney, 'Vogel's Textbook of Quantitative Chemical Analysis', Longman Scientific and Technical, New York, 1989, Chapter 4, p. 138.

4. A.J. Walder et al., Proc. Third Int. Conf. Plasma Source Mass Spectrometry, Durham, 13-18th September 1992.

5. S.R.N. Chenery and J.M. Cook, Proc. Third Int. Conf. Plasma Source Mass Spectrometry, Durham, 13-18th September 1992.

Laser Ablation ICP-MS on Spent Nuclear Fuel

J. I. García Alonso, J. García Serrano, J.-F. Babelot,
J.-C. Closset, G. Nicolaou, and L. Koch
COMMISSION OF THE EUROPEAN COMMUNITIES, JOINT RESEARCH
CENTRE, INSTITUTE FOR TRANSURANIUM ELEMENTS, P.O. BOX 2340,
7500 KARLSRUHE, GERMANY

1 INTRODUCTION

The analysis of fission products and actinides is carried out normally after dissolution of the fuel in 7M nitric acid (PUREX type conditions). U and Pu are preferentially determined by isotope dilution employing TIMS[1,] while fission products and minor actinides can be analysed by ICP-MS[2]. However, the dissolution of spent nuclear fuel in nitric acid and posterior dilutions for analysis require extensive sample preparation in a hot cell provided with master-slave manipulators. Also, the gaseous and volatile fission products (Kr, Xe, I) are lost during dissolution while an insoluble residue containing noble fission metals as well as U and Pu traces is formed.

It is clear that the elemental and isotopic analysis of spent nuclear fuels would be simplified if the solid samples could be analysed directly in a hot cell without any chemical conditioning. Furthermore, spatial information for radionuclide concentrations could be gathered which otherwise would be lost when dissolving the sample. Current methods for the spatial analysis of solid highly active samples mainly include Electron Microprobe Analysis (EMPA) and Gamma Spectrometry. Electron microprobe analysis offers high spatial resolution but lacks enough sensitivity to detect low concentrations of fission products and minor actinides and does not give isotopic information. On the other hand, gamma spectrometry can be applied only to radioactive isotopes (e.g. [137]Cs). In this context, Laser Ablation coupled to ICP-MS could offer an alternative or complementary solution for these type of studies. However, the spatial resolution would not be comparable to electron microprobe analysis. Also, quantification of fission products and actinides in spent nuclear fuel by laser ablation ICP-MS would be a difficult task due to the non-existence of available reference materials containing all fission products and actinides. The use of SIMFUEL (simulated nuclear fuel)[3] as reference could alleviate some of the problems for selected fission products but one has to bear in mind that SIMFUEL is not an elemental reference material.

A commercial uranium oxide (UO_2) fuel contains ca. 85% of uranium of which typically 3.2% is [235]U before irradiation. During irradiation 3 to 5% fission products are produced both from the original [235]U and from [239]Pu formed by

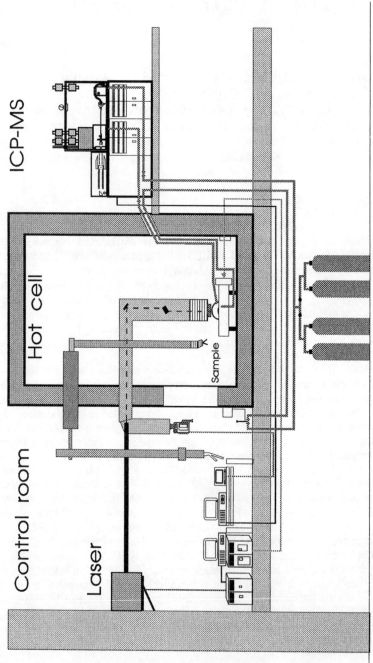

<u>Figure 1</u> Schematic diagram of the Laser Ablation ICP-MS set-up installed in the hot cell

neutron capture of ^{238}U. Plutonium and other transuranium elements (Np, Am, Cm) are also produced as a result of successive neutron capture reactions. The formation of ^{239}Pu is known to be preferential at the rim of the fuel pellet[4] due to the absorption of epi-thermal neutrons by ^{238}U in this area (which is the origin of the so-called rim effect). As a result, fission products are also enriched in the cladding-fuel boundary.

Due to the high temperature gradient between the centre and the exterior of the fuel pellet, migration of volatile fission products can also be observed (e.g. Cs). The net result of the rim and of diffusion effects is a large increase in the concentration of fission products and transuranium elements in the pellet to cladding boundary with a decrease in the grain size and the appearance of a porous microstructure[4]. If we consider the direct repository of spent fuel in deep geological formations the concentrations of fission products and actinides in the fuel rim should have to be taken into account in the computer predictions of the long-term stability of the fuel. All the effects previously described have been observed by electron microprobe analysis and microscopy[4]. However, for some low-yield fission products, like iodine, EMPA is not sensitive enough and nothing is known about spatial changes in isotopic abundances of fission products and actinides. The application of Laser Ablation ICP-MS to spent nuclear fuel could answer some of these questions.

2 EXPERIMENTAL

Instrumentation

The ICP-MS used is an Elan 250 from Sciex, Canada, which was modified in order to analyse radioactive samples in a glove box. Details of the modified ICP-MS instrument have been presented elsewhere[5]. The plasma torch and sliding interface, with the sampler and skimmer cones, are installed inside the glove box while the mass-spectrometer and associated electronics are outside. The Laser Ablation instrument is a Laser Sampler 320 from Perkin Elmer which was also modified for the analysis of radioactive samples in a hot cell.

Figure 1 shows a schematic diagram of the final set-up. The ablation stage takes place inside the hot cell (alpha, beta and gamma protection). The laser (IR, Nd-YAG) has been installed on the wall opposite the hot cell and the light beam is sent into the cell through a series of two IR mirrors and one IR lens installed in a periscope connecting the control room with the hot cell. The periscope is fitted with two optical paths, one is used for the laser and the other for a video camera to observe the sample and position it for ablation. The ablation cell was installed inside the hot cell with the original x-y-z translation stage and was modified to handle the ablation stage by master-slave manipulators. The ablated aerosol is carried with an argon flow of, typically, 1 l/min (mass flow controller) through a PVC tubing of 8 m in length, 4 mm i.d., to the ICP-MS instrument situated behind the hot cell. In the standard operating conditions the ablated aerosol takes 6 seconds to reach the plasma.

Operating conditions

The spent fuel pellet was cut and embedded in Araldite and then polished. The uranium oxide fuel diameter was 9.1 mm and surrounded by a 1 mm thick Zircalloy cladding. Sampling was performed along the diameter of the pellet including the cladding every 0.5 mm so, approximately, 20 measurements were performed for spatial studies.

The laser is operated in Q-switch mode at 65 mJ per pulse and a repetition rate of 4 Hz. A flow of 1 l/min Ar carries the ablated aerosol to the plasma operating at 1400 watt forward power. For a qualitative examination of the fuel a 5 minutes ablation is performed on one same spot and the data are collected in scanning mode in high resolution (20 measurements per peak). For spatial studies the data are collected in peak jumping mode at low resolution (1 measurement per peak). In this case, the total ablation time is 60 seconds while the first 40 seconds are used for pre-ablation and then 5 scans are acquired in the remaining 20 seconds. Other operating conditions used are summarised in Table 1.

Table 1 Operating conditions

RF power	1400 watt
Reflected power	<5 watt
Argon cooling	12 l/min
Argon auxiliary	1.4 l/min
Argon carrier	1 l/min
Load coil/Sampler cone distance	25 mm (fixed)
Interface pressure	2 Torr
Quadrupole working pressure	2×10^{-6} Torr
Sampler cone	Nickel, 1 mm orifice
Skimmer cone	Nickel, 0.7 mm orifice
Laser power	65 mJ
Repetition rate	4 Hz
Dwell time (spatial studies)	20 ms
Integration time (spatial studies)	200 ms/amu
Integration time (qualitative)	2 s/amu (20 meas.)
Total scan time (spatial studies)	2.5 s
Mass range (qualitative)	80-160 and 230-245
Mass range (spatial studies)	91-100,119,127-150,234-237,239-244,254

Peaks at mass 119 ($^{238}U^{2+}$) and 254 ($^{238}U^{16}O^{+}$) were measured and used as internal standards as the concentration of ^{238}U varies only slightly along the pellet diameter.

3 RESULTS AND DISCUSSION

Characterisation of spent fuel

In Fig. 2 the intensity vs time profile for four selected isotopes in spent fuel is shown. The total ablation time was 2 minutes. As can be observed adequate signals can be obtained both for actinides and fission products. [136]Xe, which is not present in the pellet surface, appears later than the other isotopes once the external layers of the fuel have been ablated. No important memory effects have been observed and the signal decays to background levels within 1 minute after finishing the ablation. In some cases, residual Xe peaks have been detected even 5 minutes after ablation which can be used to study fission Xe isotopic abundances in the absence of Cs and Ba isobaric interferences.

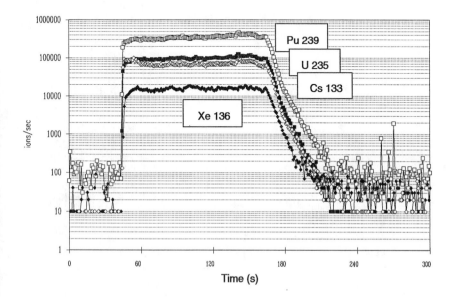

<u>Figure 2</u> Intensity vs time profile for 2 minutes ablation of spent nuclear fuel

The fission product distribution of the spent fuel as detected with the laser ablation system is shown in Fig. 3. Except for Kr (masses 83, 84, 85 and 86) all other fission products can be detected. Peaks at masses 85 and 87 correspond to Rb, 88 and 90 to Sr, 89 to Y, 90, 91, 92, 93, 94 and 96 to Zr, 95, 96, 97, 98 and 100 to Mo, 99 to Tc, 100, 101, 102 and 104 to Ru, 103 to Rh, 104, 105, 106, 107, 108 and 110 to Pd, 109 to Ag, 110, 111, 112, 113 and 114 to Cd. The peak at mass 119 corresponds to $^{238}U^{2+}$ which shows the symmetry of the fission process. For the other half of the fission products, peaks at masses 126, 128 and 130 correspond to Te, 127 and 129 to I, 131, 132, 134 and 136 to Xe, 133, 134, 135 and 137 to Cs, 134, 136, 137 and 138 to Ba, 139 to La, 140, 142 and 144 to Ce, 141 to Pr, 142, 143, 144, 145, 146, 148, 150 to Nd, 147 to Pm, 147, 148, 149, 150, 151, 152 and 154 to Sm, 151, 152, 153, 154 and 155 to Eu and 154, 155, 156, 157, 158 and 160 to Gd. As can be observed some of the peaks can be assigned to more than one element due to beta decay or neutron capture reactions during fuel irradiation. Also peaks at masses 154 to 160 could be attributed partially to the oxides of Ba, La, Ce and Nd. When the spent fuel is dissolved in nitric acid for analysis, all I and Xe peaks disappear and the peaks for Zr, Mo, Tc, Ru, Rh and Pd are reduced significantly[5] as they partially form insoluble residues. In this context laser ablation could produce a more complete picture of the fission product inventory. The lack of reference materials, however, prevents the acquisition of quantitative results. The use of simulated spent fuel (SIMFUEL)[3] for calibration will be tested for future studies.

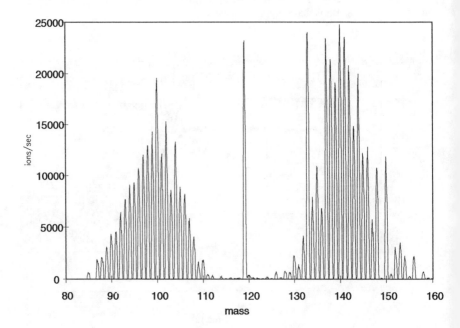

Figure 3 Fission product distribution of spent nuclear fuel obtained by laser ablation sampling

Spatial distribution of fission products and actinides

For the study of the spatial distribution of fission products and actinides in spent fuel, 20 points along the pellet diameter, including the cladding, were measured under the conditions described above. Five scans of the mass ranges 91-100, 119, 127-150, 234-237, 239-244 and 254 were obtained in each point and the ratios to the peaks at masses 119 ($^{238}U^{2+}$) and 254 ($^{238}U^{16}O^+$) used to study relative concentration changes. No other internal standard can be used for this type of sample. The radial distribution of uranium, neptunium and plutonium isotopes normalized to the $^{238}U^{16}O^+$ peak are shown in Fig. 4. As can be observed the concentration of ^{235}U (Figure 4a) decreases gradually approaching the rim of the pellet while the concentration of ^{236}U (formed by neutron capture of ^{235}U) increases at the pellet rim. The ^{235}U to ^{236}U isotopic ratio varies from 0.62 at the pellet rim to 0.78 at the centre which shows a symmetrical non-homogeneous distribution of isotopic abundances. ^{237}Np, formed from ^{236}U by neutron capture, is also enriched at the rim. Similar results have been observed for Pu isotopes in Figure 4b. The "rim effect" is clearly observed as all Pu isotopes are formed by successive neutron capture from ^{238}U. The large increase of ^{239}Pu concentration in the rim of the fuel would give rise also to an increase in the concentration of the fission products (see below).

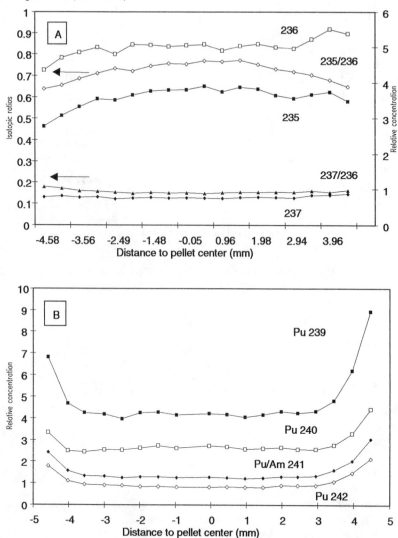

Figure 4 Relative concentrations of U, Np and Pu isotopes referred to the ^{238}U^{16}O^{+} peak. a) U and Np isotopes including isotopic ratios. b) Pu isotopes

Figure 5 shows the Pu isotopic ratios observed as a function of the distance to the pellet centre. The peak at mass 241 corresponds both to ^{241}Pu and its beta decay product ^{241}Am. As can be observed ^{239}Pu has a constant isotopic abundance with respect to the sum of all other Pu isotopes while ^{240}Pu is reduced at the pellet exterior where ^{241}Pu and ^{242}Pu increase.

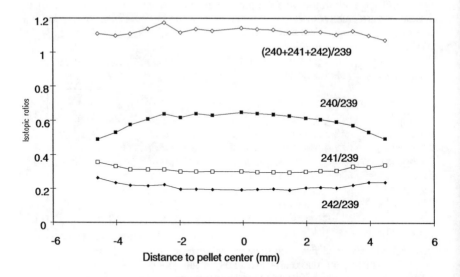

<u>Figure 5</u> Isotopic ratios for Pu isotopes with the distance to the centre of the pellet

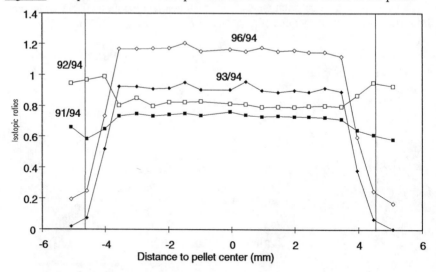

<u>Figure 6</u> Isotopic ratios for Zr isotopes with the distance to the pellet centre including the cladding containing natural Zr (vertical lines show the cladding-fuel boundary).

On the other hand, no changes in isotopic abundances for fission products have been observed across the pellet diameter. The isotopic ratios obtained for Zr are shown in Fig. 6. The natural abundances of Zr in the cladding allows to distinguish between cladding and fuel. Also, the fission yields for Zr isotopes from

235U and 239Pu are clearly different. As can be observed there is a big change in the Zr isotopic abundances at the pellet edges but no changes are observed inside. These constant isotopic ratios were observed also for Mo,Te, I, Cs, Xe, Ce and Nd, i.e. all multi-isotopic elements measured. The conclusion of these findings is that the fission yield contribution of 239Pu and 235U should be independent of the distance to the pellet centre outside of the rim zone where isotopic ratios can not be determined precisely due to our wider crater diameter.

The tabulated cumulative fission yields for selected Zr and Nd isotopes obtained from 235U and 239Pu fission[6] are given in Table 2. In order to compensate for neutron capture reactions in the reactor (e.g. formation of 92Zr from 91Zr), composite isotopic ratios are used to study fission yield contributions. The observed isotopic ratios for the Zr 91+92/93+94 and Nd 143+144/145+146 are illustrated in Fig. 7 and the mean values for the fuel introduced in Table 2. As can be observed the isotopic ratios for the fuel are constant and between the corresponding 235U and 239Pu fission. Based on the Zr ratios, it can be calculated that ca 45% of the fission comes from 239Pu and 55% from 235U . The Nd ratios can not be used for these calculations because of the similar fission yields from both nuclides. The computer code KORIGEN[7] was also used to calculate isotopic ratios for Zr and Nd given the irradiation history of the fuel, the final burn-up (53 GWd/tU), the reactor type (PWR) and its neutron spectrum. The fission yield contribution calculated from the KORIGEN results was 51% for 235U and 49% for 239Pu. As can be observed, there is a good agreement between the experimental data and the theoretical calculations for fission yield contributions of 235U and 239Pu.

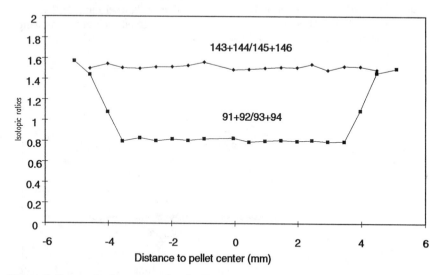

Figure 7 Composite isotopic ratios for Zr and Nd isotopes used to calculate fission yield contributions.

Table 2 Comparison between the experimental isotopic ratios obtained and those predicted by the cumulative fission yield data[6] of ^{235}U and ^{239}Pu

Isotope or isotopic ratio	Cumulative fission yield (%) from		Ratio found	KORIGEN
	^{235}U	^{239}Pu		
91 Zr	5.9174	2.4317	-	-
92 Zr	5.9924	3.9266	-	-
93 Zr	6.4128	3.7796	-	-
94 Zr	6.4539	4.2960	-	-
143 Nd	5.9933	4.4586	-	-
144 Nd	5.4492	3.7746	-	-
145 Nd	3.9534	3.0160	-	-
146 Nd	2.9989	2.4807	-	-
Zr 91+92/93+94	0.9256	0.6635	0.8078	0.8221
Nd 143+144/145+146	1.6458	1.4978	1.5118	1.5749

Finally, relative concentration changes for fission products were determined with respect to the ^{238}U^{2+} peak (mass 119). In all cases enrichment at the pellet rim was observed as illustrated in Fig. 8 for selected Mo, Tc, Ba, Ce and Nd isotopes. These elements give an indication of the relative fuel burn-up as they do not migrate due to the temperature gradient. As can be observed, a constant concentration is obtained in the centre of the pellet with increasing concentrations at both sides due to the rim effect. The width of the rim has been observed to be only of about 100 to 200 μm by EMPA[4,] depending of the fuel burn-up. However, in our results this area is not well defined and the observed concentration changes are smoothed by our wider crater diameter of about 0.5 mm.

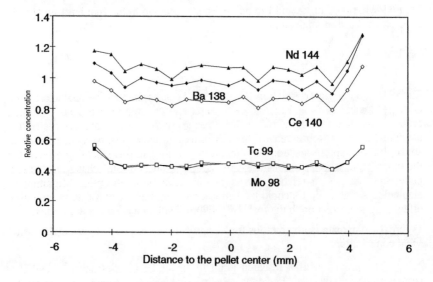

Figure 8 Relative concentration changes of non-migrating fission products in spent fuel.

Elements like Cs and I do migrate due to the temperature gradient in the fuel pellet. In order to study migration effects the ratio to a non-migrating element, such as Nd, will give an indication of the magnitude of this effect. The relative concentration changes of ^{135}Cs and ^{129}I relative to ^{144}Nd are shown in Fig. 9. As can be observed, Cs and I migrate to the pellet rim. The migration of readily soluble elements, like Cs and I, having long-lived beta emitting isotopes, such as those illustrated in Fig. 9, should be kept in mind in planning studies for the direct repository of spent fuel.

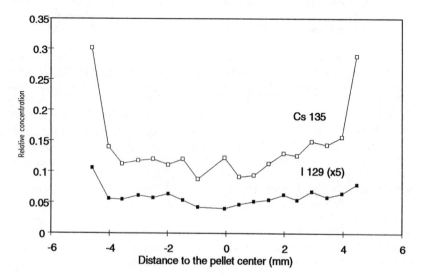

Figure 9 Migration of Cs and I in the fuel pellet relative to the fuel burn-up (^{144}Nd used as fuel burn-up indicator)

4 CONCLUSIONS

The technique of Laser Ablation ICP-MS has been applied to the characterisation of spent nuclear fuels and to study elemental and isotopic distributions of fission products and actinides in the fuel pellet. Changes in the isotopic abundances for the actinides have been observed for the first time and these results could be used to study neutron capture reactions in relation to the rim and self-shielding effects. On the other hand, isotopic abundances for fission products do not change in the fuel pellet but changes in relative concentrations have been observed. Migration of elements such as I and Cs could be demonstrated but the extent of these migration effects and their relation to the fuel temperature gradient have yet to be studied.

Acknowledgments

The provision of a research grant to Javier Garcia Serrano by ENRESA (Empresa Nacional de Residuos Radioactivos S.A., Spain) is gratefully acknowledged.

5 REFERENCES

1. M. Wantschik, B. Ganser and L. Koch, Int. J. Mass Spectr. Ion Physics, 48(1983)405
2. J.I. García Alonso, J.-F. Babelot, J.-P. Glatz, O. Cromboom and L. Koch. Radiochim. Acta, in press.
3. P.G. Lucuta, R.A. Verrall, Hj. Matzke and B.J. Palmer, J. Nucl. Mater., 178(1991)48
4. Hj. Matzke, J. Nucl. Mater., 189(1992)141
5. L. Koch, R. de Meester, S. Franzini and H. Wiesmann, "Adaptation of ICP-MS to a glove box for the analysis of highly radioactive samples". Paper presented at the "1st Int. Conference on Plasma Source Mass Spectrometry". Durham(UK), Sep. 1988
6. M.E. Meek and B.F. Rider, "Compilation of Fission Product Yields". General Electric Company. Vallecitos Nuclear Center. Pleasanton, California (1972)
7. U. Fischer and H.W. Wiese, "Verbesserte konsistente Berechnung des nuklearen Inventars abgebrannter DWR-Brennstoffe auf der Basis von Zell-Abbrand-Verfahren mit KORIGEN". Kernforschungszentrum GmbH, Karlsruhe. Report 3014(1983)

Analysis of Highly Radioactive Liquid Samples by ICP-MS

M. Betti, J. I. García Alonso, P. Arboré, and L. Koch
COMMISSION OF THE EUROPEAN COMMUNITIES, JOINT RESEARCH
CENTRE, INSTITUTE FOR TRANSURANIUM ELEMENTS, P.O. BOX 2340,
7500 KARLSRUHE, GERMANY

T. Sato
NUCLEAR CHEMISTRY LABORATORY, DEPARTMENT OF CHEMISTRY,
JAPAN ATOMIC ENERGY RESEARCH INSTITUTE, 319-11 TOKAI-MURA,
IBARAKI-KEN, JAPAN

1 INTRODUCTION

During the last decade the analytical technique of
Inductively Coupled Plasma Mass Spectrometry (ICP-MS) has
been extensively applied in different fields of research,
such as clinical chemistry, biochemistry, geochemistry as
well as environmental studies[1-4].

On account of its high sensitivity and multi-
isotopic capabilities the technique of ICP-MS has also
been employed for the characterization of spent nuclear
fuels[5].

Samples of relatively low activity, such as diluted
solutions arising from the dissolution of spent fuels,
can be handled in a glove-box and both elemental and
isotopic information can be quickly gathered for most
of the fission products and actinides. However, the
methodologies applied for the analysis of nuclear samples
by ICP-MS differ from those employed for natural
elements. This is because of the different and, in many
cases, unknown isotopic abundances in nuclear samples.
From this point of view, the determination is nuclide-
specific rather than element orientated. As well as this,
isobaric interferences from neutron capture reactions,
the consequence of the irradiation of the fuel, and from
ß-decay of relatively long-lived radionuclides, occur.
Therefore, for the complete characterization of nuclear
material the analysis must be preceded by chemical
separation, which can be time-consuming and dangerous to
health.

In this paper the on-line coupling of a
chromatographic system with an ICP-MS installed in a glove
box is described. Examples of the analysis of heavy

fission products (Cs, Ba, lanthanides), separated on-line by Ion Chomatography are presented.

2 EXPERIMENTAL

Instrumentation

An Elan 5000 ICP-MS instrument (Sciex, Canada) and a 4500i high pressure chromatographic pump (Dionex, USA), modified in order to handle radioactive materials in a glove-box, were used.

The diagram in Fig. 1, shows the installation of the system in the glove-box. The quadrupole mass filter, ion lenses and channel electron-multiplier detector are outside the glove box. All vacuum line connections to the two rotary pumps are provided with absolute filters to avoid radioactive contamination in the pumps. The mass-spectrometer is connected to the plasma through an opening in the left-side stainless-steel wall of the glove-box where the sampling interface is attached. The plasma torch, nebulizer, chromatographic column, pneumatic injector valve, peristaltic pump and autosampler are inside the glove box. The eluent, coming from the chromatographic pump outside the glove box, is passed into the injection valve and then into the chromatographic column. If no chromatographic separation has to be performed, the sample is sucked through the peristaltic pump directly into the nebulizer. When chromatographic separations are carried out before the ICP-MS detection, the effluent from the chromatographic column is directly injected in the nebulizer. In both cases the waste from the nebulizer is stored in a vessel inside the glove box.

A mixed-bed ion-exchange column CS5 (Dionex, USA) and a 100 μl loop were employed.

Reagents

Stock solutions of natural elements were obtained from Spex as 1000 ppm standards and diluted daily as necessary with 1% nitric acid. Nitric acid, oxalic acid, diglycolic acid and lithium hydroxide (Merck, Suprapur grade) as well as ultra pure water (18 MegaOhm/cm resistivity at 25°C) obtained by treating Milli-Q water in a UHQ system (Elga, U.K) were used throughout.

Eluent solutions

The eluent solutions consisted of oxalic acid and diglycolic acid both 100 mM and dissolved in 190 mM lithium hydroxide.

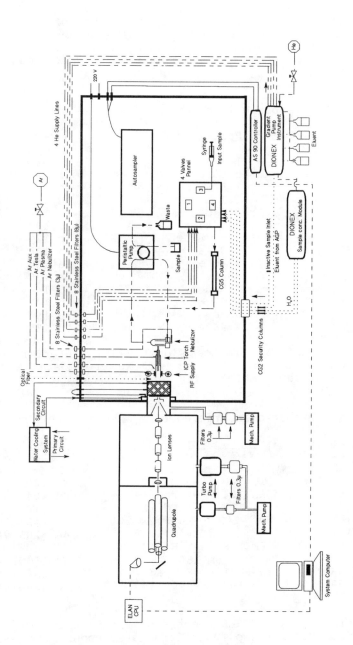

Figure 1 Installation of ICP–MS in a glove–box

3 RESULTS AND DISCUSSION

The complete determination of the abundance of fission products in spent fuels by ICP-MS is hindered by isobaric interferences of nucleides originating from ß-decay and neutron capture reactions, respectively summarized in Tables 1 and 2.

Table 1   Isobaric interferences by Beta-decay

Source isotope	Half-life	Decay product
Kr 85	10.76 a	Rb 85
*Sr 90	28.5 a	Zr 90
Zr 93	1.5 E6 a	Nb 93
Tc 99	2.1 E5 a	Ru 99
*Ru 106	368 d	Pd 106
Pd 107	6.5 E6 a	Ag 107
I 129	1.57 E7 a	Xe 129
*Cs 134	2.06 a	Ba 134
Cs 135	2 E6 a	Ba 135
*Cs 137	30.17 a	Ba 137
*Ce 144	285 d	Nd 144
*Pm 147	2.62 a	Sm 147
Sm 151	93 a	Eu 151
*Eu 155	4.96 a	Gd 155

*Isotopes which will produce significant   interference after 5 years cooling time (Kr and Xe are lost during dissolution of the sample)

Table 2 Isobaric interferences by neutron capture

Source isotope	Isotope produced	Interference on
Y 89	Y 90 ---> Zr 90	Sr 90 (28.5 a)
Mo 95	Mo 96	Zr 96
Tc 99 (2.1 E5 a)	Tc 100 ---> Ru 100	Mo 100
Rh 103	Rh 104 ---> Pd 104	Ru 104
Ag 109	Ag 110 ---> Cd 110	Pd 110
Pr 141	Pr 142 ---> Nd 142	Ce 142
Nd 143	Nd 144	Ce 144 (285 d)
Pm 147 (2.62 a)	Pm 148 ---> Sm 148	Nd 148
Sm 149	Sm 150	Nd 150
Eu 153	Eu 154 (8.8 a)	Sm 154
Tb 159	Tb 160 ---> Dy 160	Gd 160

In Fig. 2, two ICP-MS spectra of the distribution of fission products for two fuels irradiated under different conditions are shown. For both spectra, the peak located at mass 119, from U^{2+}, is common. Greater differences are evident for the high mass range (Cs, Ba and lanthanides), where the occurrence of relatively long-lived beta emitters is coupled with high neutron-capture cross-sections. The net result is the presence of isotopes of different elements at the same nominal mass (e.g. Cs-137 and Ba -137 for the beta-decay case and Nd-148 and Sm-148 for the case of neutron capture reactions).

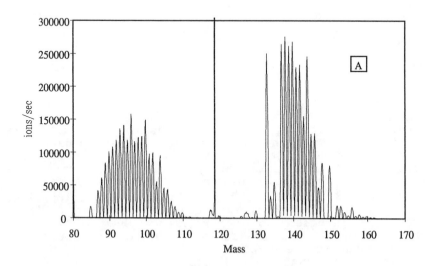

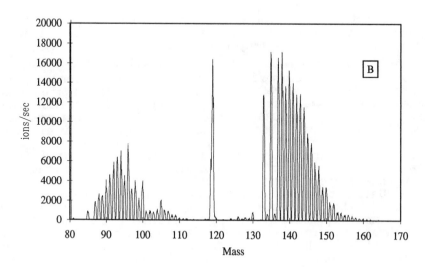

Figure 2 ICP-MS spectra of fission products in spent nuclear fuel. a) Uranium Oxide irradiated in a thermal reactor. b) Uranium-Neptunium Oxide irradiated in a fast neutron reactor.

Neutron capture reactions are absent in Fig. 2B as can be observed for the quasi-gaussian distribution of the high mass fission products. However, commercial fuels correspond typically to Fig. 2A

In order to obtained a complete picture of the fission product inventory, for characterization purposes, a chemical separation of the elements has to precede the ICP-MS measurement. Elements to be separated from each other include Cs from Ba, Ce from Nd, Nd from Sm, Pm from Sm, Sm from Eu and Sm from Gd. For this purpose an ion chromatograph has been coupled with the ICP-MS mounted in a glove-box.

For the separation, carried out in the glove-box, an agglomerated ion-exchange resin, Dionex CS5, has been used. This type of resin contains an internal core particle, consisting of polystyrene-divinylbenzene of moderate cross-linking (2-5 %), to which is attached a monolayer of small diameter particles which carry the functional groups comprising the fixed ions of the ion-exchange[6]. Provided the outer layer of functionalized particles is very thin, the agglomerated resin exhibits excellent chromatographic performance due to the very short diffusion path available to solute ions during the ion-exchange process. The CS5 column, used in this investigation, consists of anion- and cation-exchange sites both of which are available during the elution.

For the separation of the lanthanide series, using a pure cation-exchanger such as the CS3 column (Dionex, USA) together with α-hydroxyisobutyric acid (HIBA) as eluent medium, the elution sequence is from Lu^{+3} to La^{+3}. Under these experimental conditions[7], the elements of interest in our investigation, elute after 12 min. and are also not well resolved. Employing a mixed-bed ion-exchanger, such as CS5, and eluent media forming anionic complexes with the lanthanides, the sequence of elution is reversed compared with the elution performed by a pure cation-exchanger. Also Cs and Ba can be separated from each other and from the lanthanide series by cation-exchange process on CS5 column.

Using the chromatographic conditions reported in Table 3, the chromatographic separation obtained for the elements considered here, before their ICP-MS measurement, is shown in Fig. 3. The lanthanides tested, are eluted in the sequence La, Ce, Pr, Nd, Sm, with Cs in between Ce and Pr. Increasing the ionic strength of the eluent - using for instance 700 mM nitric acid - Ba is also eluted at the end of the chromatogram.

In another experiment it was confirmed that Pm, which was not present in the sample of Fig. 3, elutes in between Nd and Sm.

<u>Table 3</u> Chromatographic conditions

Column:	HPIC-CS5, 250 x 50 mm
Loop volume:	100 μl
Flow rate:	1 ml/min
Eluent 1:	100 mM Oxalic acid, 190 mM Lithium hydroxide
Eluent 2:	100 mM Diglycolic acid, 190 mM Lithium hydroxide
Eluent 3:	distilled water

Gradient program:

Time (min)	%1	%2	%3	Comment
0.0	60	30	10	start
0.5	60	30	10	injection
3.4	60	30	10	start of gradient
15.0	30	30	40	end of gradient

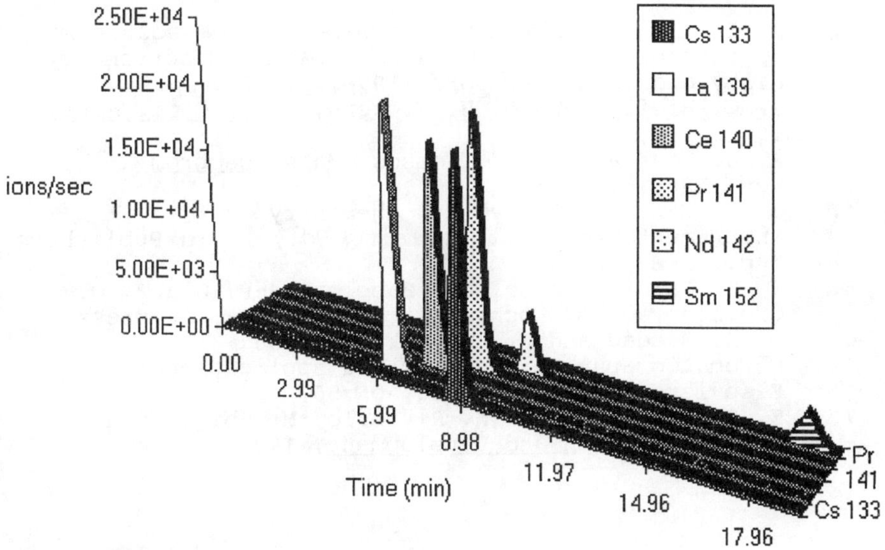

<u>Figure 3</u> Chromatographic separation of six elements with ICP-MS detection. Separation performed according to the conditions of Table 3

4 CONCLUSIONS

After having achieved the chemical chromatographic separation by the procedure here described, by taking advantage of the different isotopic abundances of fission elements to the corresponding natural ones, the methodology of isotopic dilution analysis by spiking the sample with natural elements can be applied. Except for La, Pr and Pm, all fission products from Cs to Gd can be determined by this way. La and Pr can be determined by calibration and/or standard additions and Pm by using the response curve of the instrument for the other lanthanides, as no standards are available. Using the above described technique, the direct determination - for instance - of fuel burn-up by ICP-MS based on Nd-148 is now possible without interference from Sm-148.

5 ACKNOWLEDGEMENTS

We acknowledge very much the workshop staff, Mr. Dockendorf, Mr. Schrodt and Mr. Ougier for their fruitful technical support in setting the instrumentation in glove-box.

6 REFERENCES

1.   A. L. Gray, 'Inductively coupled plasma source mass spectrometry' in 'Inorganic Mass Spectrometry', Wiley and sons, New York, 1998, p. 257.
2.   G. M. Hieftje and G. H. Vickers, Anal. Chim. Acta, 1989, 216, 1.
3.   R. S. Houk and J. J. Thomsos, Mass Spectrom. Rev.,1988, 7, 425.
4.   J. E. Cantle, 'Analytical chemistry Instrumentation', W. R. laing Ed., Lewis Publishers Inc., 1986, p. 3.
5.   J. I. Garcia Alonso, J-F Babelot, J-P. Glatz, O. Cromboom and L. Koch, Radichimica Acta, in press
6.   P. R. Haddad and P. E. Jackson, 'Ion Chromatography, principles and applications', Elsevier, Amsterdam, 1990, p. 56.
7.   S. S. Heberling, J. M. Riviello, M. Shigen and A. W. Ip, Research and Development, 1987, 9, 74.

Impurity Assay in Reprocessing Uranyl Nitrate

P. Leprovost, R. Schott, and A. Vian
COGEMA, LA HAGUE, SERVICE LABORATOIRE, F50444 BEAUMONT-HAGUE, CEDEX, FRANCE

1 INTRODUCTION

ICP/MS appears as a powerful analytical tool in Nuclear Industry. It enables the chemist to assay impurities in products such as Uranyl Nitrate and Plutonium Dioxide. Isotopic ratio measurements on U and Pu are possible too. Trace element determinations in the environment have been performed with the same apparatus.

This lecture will show how ICP/MS helps our Laboratory to match new requirements related to :

* productivity: increasing production rate of recent reprocessing units implies an increased agenda for the specification checking analyses.

* analytical efficiency: the number of elements to check tends to increase and the concentration levels to be lowered.

* drastic limitation of organic wastes: their removal from analytical boxes is a serious probem; Safety Organisations no longer admit the mixing of solvents with plaster and their disposal as solid wastes.

The product specification checking classically requires the assay of 38 elements beside the quantitation of some anionic species, the measurement of physical and radiometric characteristics. The elemental assays demand several solvent extractions of uranium or plutonium before applying ICP/Atomic Emission Spectrometry or flame Atomic Absorption Spectrophotometry.

The new technique combines sensitivity, selectivity, quickness and near zero production of organic wastes. It thus allows us to perform determinations without uranium removal, with good precision and quite satisfactory detection limits.

2   INSTRUMENTATION

The ICP/MS instrument is a VG PlasmaQuad (fig.1). The operating conditions are shown in table 1.

Table 1 Operating conditions

PLASMA TORCH	
* Power	1300 W
* Reflected power	<10 W
* Plasma gas flow	13 l/mn
* Auxiliary gas flow	0.7 l/mn
* Carrier gas flow	0.8 l/mn
NEBULISER	
* Type	Meinhardt
* Solution flow rate	0.3 ml/mn
* temperature of the spray chamber	12°C
SPECTROMETER	
* Interface pressure	2.7 mbar
* Lens compartment pressure	$<10^{-4}$ mbar
* Quadrupole compartment pressure	5×10^{-6} mbar

3   PROCEDURE

The analytical procedure is to be called the "semi-quantitative method", which includes the use of the "mass response" curve.

Recording the "Mass Response" Curve

A standard solution containing equal concentrations of 13 elements (table 2) is analysed with the instrument. The elements have been chosen in order to cover the mass range to scan. Indium is used as an internal standard.

Table 2 Elements used for the"Mass-Response" Curve

Li	Be	Mg	V	Co	Y	Pd	In	Pr	Eu	Dy	Pb	Th
7	9	24	51	59	89	106	115	141	153	164	208	232

Figure 1

SCHEMATICS OF ICP/MS

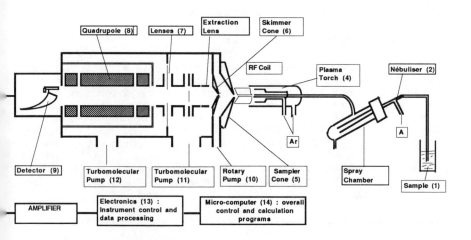

Figure 2

MASS RESPONSE CURVE

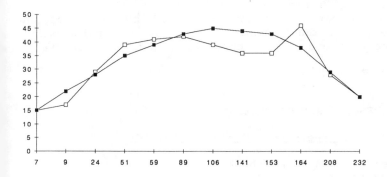

The response curve is adjusted on the dots representative of the ratio **r** vs **A** (the atomic mass):

$$r_X = N_X/N_{In} \qquad\qquad\qquad (1)$$

$$N_X = \text{counts for isotope } A_X$$

$$N_{In} = \text{counts for } {}^{115}In$$

The curve is bell-shaped (fig.2), with a maximum about mass 120. The curve includes all instrumental effects but not the ionization efficiency which may differ from one element to another; this last parameter explains the wide scattering of the dots around the adjusted curve. The correction for ionization efficiency or "Saha factor" correction is applied to the elements of interest:

* the 13 elements for the mass response calibration,
* the other elements to be assayed, if they differ from the previous.

Saha factor correction

Let $[X]_0$ be the concentration of element X in the standard solution and $[X']_0$ the concentration found when the standard solution is analysed as an unknown. The factor for correcting the ionization effect is :

$$k_X = [X]_0/[X']_0$$

Analysis

The mass spectrum of an unknown solution containing the elements to be assayed and Indium with known concentration is recorded. The ratio r_X (1) is converted into concentration of isotope $[A_X]$ by means of the response curve. The isotopic abundancy (a) of isotope A_X being known, the elemental concentration may be computed:

$$[X] = [A_X]/a$$

4   RESULTS AND DISCUSSION

Evaluation of precision and accuracy (table 3)

The elements to be checked are added to a solution of reprocessing uranyl nitrate; the concentration is 1.4 µg/g U for most of the elements except for silver, calcium (6.9 µg/g U), iron and silicon (69 µg/g U). The solution is diluted with 1 M nitric acid (factor 500) and analysed 10 times, using the unspiked solution as a blank.

The relative standard deviation for repeatability (RSD1) is generally between 3 and 15% ; the greatest discrepancy is observed for:

* the low mass elements (e.g. Lithium), for which the quadrupole monitoring is less stable;

* the elements sensitive to external pollution (Ca, Mg, Zn);

* silicon, subject to a strong polyatomic interference by $^{14}N_2^+$ at mass 28.

Ten solutions identical to that used to evaluate RSD1 were prepared and analysed in the same way. The relative standard deviation (RSD2) is generally smaller than 15%. The overall reproducibility is not significantly different from the repeatability on a single solution; the difficulties are encountered for the same elements.

The accuracy for each element is evaluated by the relative difference $\Delta\%$ between the measured concentration and the expected value.

Evaluation of detection limits (table 3)

Detection limits are below 1 μg/g U except for:

- silver (1.3 μg/g U),
- calcium (7 μg/g U); strong fluctuations of the blank due to external pollutions,
- silicon (79 μg/g U); polyatomic interference by $^{14}N_2^+$ at mass 28,
- iron (29 μg/g U), polyatomic interference by $^{40}ArOH^+$ at mass 57,

Table 3   Uncertainty - Detection limits

Element	Concen-tration Level	RSD1	RSD2	Δ	Detection Limit
		Repeat ability	Reproduci bility	Accuracy	(4 sigma)
	μg/g U	%	%	%	μg/g U
Li	1.4	9	10	15	0.48
Be	1.4	15	15	10	0.80
B	1.4	10	12	3	0.68
Na	1.4	11	23	27	0.44
Mg	1.4	15	22	21	0.32
Al	1.4	9	11	16	0.36
Si	69	23	18	33	79
Ca	6.9	30	31	5	7
Ti	1.4	6	4	8	0.28
V	1.4	5	5	-2	0.28
Cr	1.4	7	7	5	0.28
Mn	1.4	3	6	7	0.16
Fe	69	10	8	14	29
Co	1.4	6	5	1	0.28
Ni	1.4	10	12	1	0.36
Cu	1.4	8	17	11	0.24
Zn	1.4	29	32	-1	0.68
As	1.4	11	9	1	0.44
Zr	1.4	3	5	-1	0.20
Nb	1.4	3	4	-1	0.20
Mo	1.4	8	9	-4	0.44
Ag	6.9	6	6	-8	1.26
Ru	1.4	5	5	1	0.24
Cd	1.4	8	13	-4	0.40
Sn	1.4	15	16	-10	0.40
Sb	1.4	9	16	0	0.36
Ba	1.4	5	12	17	0.24
Ta	1.4	7	5	-5	0.92
W	1.4	8	7	-4	0.48
Pb	1.4	6	6	4	0.28
Bi	1.4	7	4	0	0.28
Sm	1.4	8	13	-1	0.36
Eu	1.4	6	6	-5	0.28
Gd	1.4	5	8	2	0.28
Dy	1.4	7	5	-6	0.40
Th	1.4	7	6	-4	0.40

Analysis of uranyl nitrate batches. (table 4)

In 1991, 21 batches of uranyl nitrate were analysed by classical methods ; these have been re-analysed with ICP/MS. The latter results confirm the former with improved detection limits. The exceptions are for:

- Li, Be and B; the limits nevertheless remain of the same order of magnitude,

- Si and Fe ; the loss of performance is greater but the limits are low enough to allow specification checking.

We must also note that K and P are not reached by our ICP/MS apparatus for these elements are too strongly interfered respectively by:
- $^{38}ArH^+$ at mass 39,
- NOH^+ at mass 31.

Graphite furnace atomic absorption is a good alternative method we apply for assaying Si, Fe, K and P (with a discharge lamp for this last element) without removing uranium. We thus need no longer use solvent extractions to assay impurities in uranyl nitrate.

The mass response curve method

The method using a calibration graph for each element is theoretically more precise than the global mass response curve. But for determinations close to detection limits, it is not necessary to record numerous calibration lines with several standards; this procedure would be time consuming with poor gain because ultimate precision is not the first purpose of the analysis.

Reducing the number of solutions analysed brings some advantages:

- the consumption of nuclear material is reduced, which limits pump contamination; this will be valuable for plutonium analysis;

- the economic gain is quite noticeable; the analysis takes only 4 hours instead 21 hours formerly; the overall duration of specification checking is shortened by 30% for a total of 56 hours.

The last advantage of the response curve method is specific to nuclear industry: one manages more easily the changes of isotopic abundancies, which depend on the fuel being reprocessed. Actually, it is not necessary to calibrate and perform analysis with the same elements or isotopes. Calibration is done with naturally occurring elements and different data bases may be used for analysis; these data bases are derived from reactor data.

<u>Table 4</u> Results obtained with actual UN batches

Element	Specification µg/g U	AES-AAS µg/g U		ICP-MS µg/g U	
Li	eq B,F⁻ non vol.	<	0,3	<	0.48
Be	F⁻ non vol.	<	0,2	<	0.8
B	<1 + eq B	<	0,2	<	0.68
Na	< 10		6	<	0.44
Mg	< 10	<	1	<	0.32
Al	F⁻ non vol.	<	2	<	0.36
Si	Si+P<500	<	5	<	147
Ca	< 15		4	<	7
Ti	< 1	<	1	<	0.28
V	< 2	<	1	<	0.28
Cr	< 15	<	2	<	0.28
Mn	F⁻ non vol.	<	1	<	0.16
Fe	< 300	<	3	<	72
Co	eq B	<	2	<	0.28
Ni	< 100	<	3	<	0.36
Cu	F⁻ non vol.	<	1,5	<	0.24
Zn	F⁻ non vol.	<	2	<	0.68
As	< 2	<	2	<	0.44
Zr	F⁻ non vol.	<	10	<	0.2
Nb	< 1	<	1	<	0.2
Mo	< 2	<	1	<	0.44
Ru	< 1	<	1	<	0.24
Ag	F⁻ non vol.	<	5	<	1.26
Cd	eq B	<	1	<	0.4
Sn	F⁻ non vol.	<	3	<	0.4
Sb	< 1	<	1	<	0.36
Ba	F⁻ non vol.	<	10	<	0.24
Ta	< 1	<	1	<	0.92
W	< 1	<	1	<	0.48
Pb	F⁻ non vol.	<	1	<	0.28
Bi	F⁻ non vol.	<	5	<	0.28
Sm	eq B	<	2,2	<	0.36
Eu	eq B	<	0,45	<	0.28
Gd	eq B	<	0,9	<	0.28
Dy	eq B	<	0,45	<	0.4
Th	F⁻ non vol.	<	20	<	0.4

5 CONCLUSION AND FURTHER DEVELOPMENT

Almost all the impurities in uranyl nitrate are now assayed by ICP/MS. The three expected gains (shorter analysis times, waste reduction and better detection limits) have actually been reached. Technological means to solve the remaining problems concerning a few elements do exist; the next part of our work will consist in using them for our purpose.

For instance we may expect serious improvements by using two new sample introduction devices:

- Electro-Thermal Vaporization (ETV) will remove nitric acid before sample injection in plasma; this will reduce matrix originated polyatomic interferences.

- With a Flow Injection Analysis System (FIAS) high salt solutions can be injected ; this device will allow smaller dilution factors, leading to better detection limits.

Finally, we hope that the results will be similar with plutonium oxide as soon as the ICP/MS instrument is installed in a glove box, in a few months.

Experimental Studies of Ion Kinetic Energies in ICP-MS

S. D. Tanner
SCIEX®, 55 GLENCAMERON ROAD, THORNHILL, ONTARIO L3T 1P2, CANADA

1 INTRODUCTION

In the conventional molecular beam interface for ICP-MS instruments, the high temperature, atmospheric pressure plasma of the ICP expands in continuum flow through a sampling orifice. The plasma expands supersonically and isentropically into a region of reduced pressure (typically 2 to 7 mbar). The isentropic core, characterized by a particle density which drops as the square of the distance from the sampling orifice and by a low neutral gas temperature, is bounded by a barrel shock and terminated at the mach disk. These shock boundaries are characterized by a rapidly increased temperature and pressure, induced by collisions of the expanding beam with the background gas species. It is now common that a skimmer is placed within the isentropic core, forming a plasma beam which feeds into the higher vacuum (ca. 10^{-3} to 10^{-5} torr) ion optical region. The gas dynamics of the expansion process have been described by Douglas and French[1].

In the expansion process, the thermal energy of the plasma species is converted to directed axial velocity. In the continuum flow through the sampler the plasma expands at constant velocity, that velocity being determined by the velocity of the bulk plasma species (Ar neutral atoms in an Ar ICP). Therefore, lighter plasma species possess a lower kinetic energy, and heavier plasma species gain a higher kinetic energy. The kinetic energy gained through the expansion by the bulk Ar neutral species, $E_{Ar,s}$ is given by:

$$E_{Ar,s} = \frac{5}{2} k T_0 \qquad (1)$$

where T_0 is the neutral plasma temperature. Of this energy, $3/2\, kT_0$ is derived from conversion of the translational degrees of freedom of the neutral species within the plasma, and kT_0 is derived from the enthalpy of the expansion. The corresponding kinetic energies of plasma species of mass M, $E_s(M)$, including ions, is given by:

$$E_s(M) = \frac{M}{M_{Ar}} E_{Ar} \qquad (2)$$

where M is the mass of the plasma species under consideration and M_{Ar} is the mass of the bulk plasma component. Combining equations (1) and (2) suggests that a plot of ion kinetic energy *vs.* ion mass should be a straight line with a slope which is proportional to the neutral plasma temperature.

The equations given above assume that the ion energy is derived solely from the supersonic expansion of the plasma. In fact, if the plasma in front of the sampler has a DC potential offset relative to ground, this DC offset will add to the supersonic energy term, yielding ion energies of the form:

$$E(M) = \frac{M}{M_{Ar}} \left(\frac{5}{2} k \, T_0 \right) + P_0 \qquad (3)$$

where P_0 is the DC plasma potential offset. In this case, the plot of ion kinetic energy *vs.* ion mass will yield as its intercept at zero mass the plasma potential offset.

Furthermore, Tanner[2] has suggested that space charge effects downstream of the skimmer cause a transmission bias towards ions of higher kinetic energy. As a result, the plot of ion kinetic energy *vs.* ion mass may show curvature at low mass (low kinetic energy). In fact, any energy-dependent focusing in the ion optics (*e.g.*, an axial stop) may skew the observed ion kinetic energies.

Ion kinetic energies have been reported for a number of ICP-MS instruments[3-7]. Various approaches have been taken to the measurement of these energies. In all instances, ion kinetic energies are obtained by scanning the DC potential of a lens element and observing the ion transmission as a function of this potential. The most common approaches include applying a retarding DC quadrupole rod bias to the analyzing quadrupole[3,4,5], variation of the lens bias potential[6], and using retarding plate analysis[7]. In all of these approaches, the data are subject to bias resulting from focusing effects.

The data presented in this work were obtained using a triple grid energy analyzer in an einzel configuration which has been described by Douglas[8]. The first and third lens elements of this analyzer serve to shield the central element from outside field penetration. By varying the potential of the central element, a stopping curve may be generated. This method minimizes ion energy dependent focusing effects as the ions exit the einzel array with the same kinetic energy with which they entered. The use of mesh lens elements permits subsequent mass analysis, thereby yielding ion energies as a function of ion mass.

From a practical point of view, a knowledge of the ion kinetic energy distribution is critical for the development of optimum ion optical systems. In addition, and this is really the focus of this paper, provided that (i) the

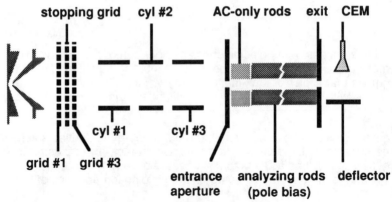

Figure 1 Ion optical path for ion kinetic energy measurements.

ion energies are "well-behaved", (ii) the measurements are relatively unbiased by focusing effects, and (iii) the gas dynamic and electrostatic influences on the ion kinetic energies are well-understood, the ion kinetic energy data can provide good diagnostic information regarding the plasma itself in front of the sampler orifice.

2 EXPERIMENTAL

The data has been obtained on a modified PE-SCIEX® ELAN® 5000 ICP-MS instrument. Modifications to the standard instrument included removing the bessel box, moving the cylinder lenses downstream and adding a triple grid energy analyzer approximately 50 mm downstream of the tip of the skimmer. A schematic of the ion optics is given in Figure 1.

The triple grid energy analyzer and the cylinder lenses were operated as einzel lens elements (that is, grid #1 and grid #3 were held at the same potential, as were cyl #1 and cyl #3). For most of the experiments, stopping curves were obtained by scanning the stopping grid. In one set of experiments, stopping curves were obtained by scanning the pole bias of the analyzing quadrupole rods. In all instances, the quadrupole was operated as a mass analyzer (peak hopping mode).

Stopping curves were obtained for a solution containing 10 ppb Li, Mg, Co, Rh, Ce, Tb, Pb and U. The experimental protocol involved setting the stopping grid to a particular potential and averaging 10 measurements for each of the analyte ions monitored. The stopping potential was then adjusted, and the measurement process repeated. The stopping potential was incremented in equal steps between 0 V and 20 V, and then was decremented in equal steps at the mid-range of the incremented steps. This process then yielded one stopping curve for each analyte ion (obtained simultaneously for all analyte ions) including data obtained both for increasing and decreasing the stopping potential (in order to minimize hysterical effects). This raw data was then subjected to a cubic natural spline smoothing and natural spline interpolation. First

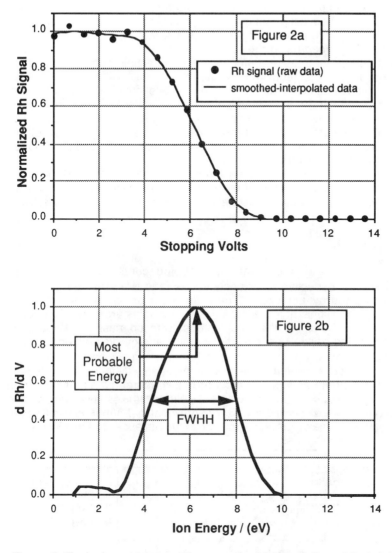

<u>Figure 2</u> Typical raw data and the smoothed-interpolation of that data are given in Figure 2a for Rh analyte ion. Figure 2b gives the first derivative of the interpolated data, which yields the ion energy distribution for Rh analyte ions under the given conditions.

differences of the interpolated data were taken to yield the ion kinetic energy distribution for each analyte ion. An example of the results is provided in Figures 2a and 2b. The points given in Figure 2a are the raw data, and the line is the cubic spline interpolation. The most probable energy for the ion is indicated in Figure 2b as is the full width at half height (FWHH) which is often used to describe the bandwidth of the energy distribution.

The results presented in this paper were obtained for varying certain plasma and ion optical parameters. Specifically, a comparison is made of ion energy measurements obtained using a triple grid energy analyzer with those obtained using the quadrupole pole bias. The impact on the measured ion energies of a change in the potentials of the electrostatic ion optical lens elements will be shown. Results will be provided for various pressures in the interface region. These latter data were obtained simply by choking the mechanical pump which evacuates the region between the sampler and skimmer. Choking this pump increases the interface pressure, which in turn induces more rapid penetration of the background gas into the isentropic core of the expansion and thus moves the mach disk forward. These experiments were performed in order to determine the impact of sampling within the mach disk. Finally, data will be presented as a function of the nebulizer gas flow, with the intent of following the temperature in the plasma through the ionization "bullet".

3 RESULTS

Figures 3 and 4 present the stopping curves and ion energy distributions, respectively, for analyte ions (and N+) obtained using the triple grid energy analyzer. The data were obtained under normal operating conditions. It is clear that the ion energies are mass dependent (the peaks of the distribution plots increase with ion mass). It will also be noticed by the astute observer that the ion energy distributions for the lightest elements (Li+, N+, Mg+) overlap; it is significant that even the background ion N+ overlaps the analyte ion energy distributions. Figure 5 presents this data as a plot of most probable energy *vs.* ion mass. The slope of the plot of Figure 5 yields a neutral plasma temperature for the conditions of the experiment of 3774K (the straight line is a fit for those ions having a mass greater than 40 amu, because of the clear non-linearity below this mass).

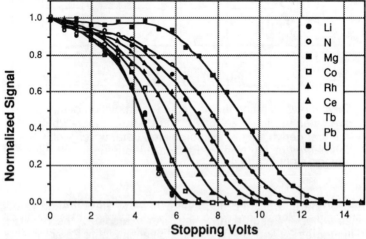

<u>Figure 3</u> Stopping curves for various analyte ions and the background ion N+. Note that the stopping curves for Li+, N+ and Mg+ overlap.

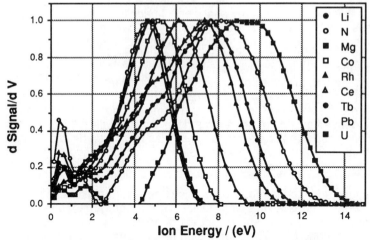

<u>Figure 4</u> Ion energy distributions derived from the data of Figure 3.

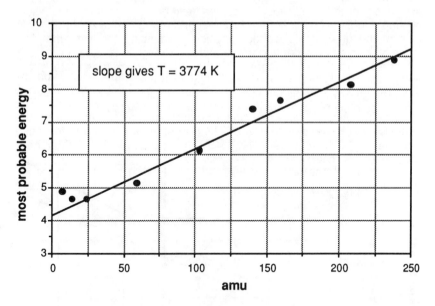

<u>Figure 5</u> Most probable ion kinetic energy *vs.* ion mass. The data was derived from Figure 4.

It should be clear that the ion energies measured are those that are appropriate only for those ions which survive through to mass analysis and detection. Therefore, any energy discrimination either before or after the triple grid energy analyzer will skew the data. This is apparent in the data of Figure 5, for which the lowest mass elements (Li+, N+ and Mg+) show essentially similar most probable energies. This result has been predicted[2] to result from space charge effects downstream of the skimmer (and probably before the energy analyzer). In addition, bias potentials of the lenses downstream of the energy analyzer can lead to biased ion energy measurements. Ion energy data has been obtained for two sets of ion energy focusing conditions. From an analytical point of view, these different lens voltages are essentially indistinguishable. However, as shown in Figure 6, the lens potentials have a significant impact on the ion energies (and hence calculated neutral plasma temperature) obtained. Although the plasma conditions were the same for the two experimental runs, the derived plasma temperatures differ substantially. As a result, it can be concluded that the plasma temperatures measured by this method are not quantitative, but for a given set of lens conditions should give data which are of qualitative value.

It was suggested in the introduction of this paper that the use of the mass analyzer pole bias as a retarding device for ion energy measurements could lead to biased ion energy measurements. In some instances (notably the work of Fulford and Douglas[5]), quadrupole pole bias measurements do indeed give a reasonably good representation of the true ion energies. It is shown here, though, that these results can not always be relied upon. Stopping curves have been measured for identical plasma conditions using either the triple grid energy analyzer or the pole bias of the quadrupole without physically adjusting the system. The stopping curves thus derived are compared in Figure 7. It is quite clear that the results do not agree. Although the pole bias data show qualitative results that agree more-or-less with the triple grid results, the methods give widely disparate quantitative results. It can be suspected that the pole bias results are skewed by focusing effects within the

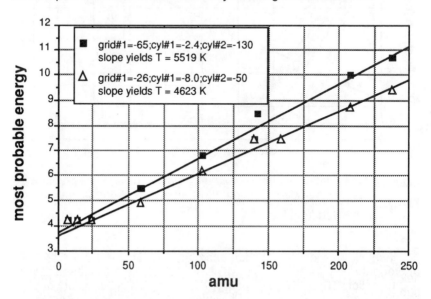

Figure 6   Most probable energy *vs.* ion mass for two different electrostatic focusing conditions.

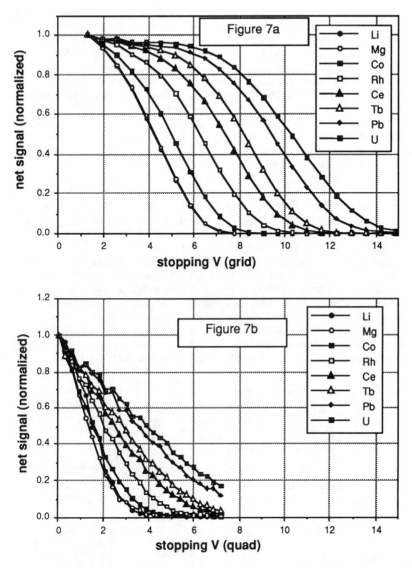

<u>Figure 7</u> Comparison of stopping curves obtained using triple grid energy analyzer (Figure 7a) and mass analyzing quadrupole pole bias (Figure 7b).

quadrupole. Such effects should be expected since for a given pole bias, lower energy (lower mass) ions spend more time both within the fringing fields of the quadrupole and within the quadrupole, and are therefore subject to significantly more cycles within the rf field than are higher energy (higher mass) ions.

As noted in the introduction, it is of some interest to study the effects of sampling from within the mach disk of the plasma expansion into the

interface region. The straightforward manner in which to do this
experiment is to simply shim the interface components until the spacing
between the sampler and skimmer is approximately the displacement of
the mach disk. However, it is simpler to increase the pressure in the
interface region, which also has the effect of moving the mach disk
forward. Under normal operating conditions (sampler diameter of 1.14
mm and an interface pressure of *ca.* 4.5 mbar), the mach disk should
form at approximately 11.5 mm downstream of the sampler. The skimmer
is normally placed a distance 6.9 mm downstream of the sampler, and so,
under these normal conditions, the skimmer should be well within the
isentropic core. If the pressure in the interface is increased to *ca.* 12
mbar, the mach disk moves up to *ca.* 6.9 mm, at which point the skimmer
is likely to be sampling from the mach disk. Ion energy measurements
were made at various interface pressures between 4.5 mbar and 10.0
mbar. The resulting stopping curves are provided in Figure 8. It is clear
that as the pressure is increased the mass dependence of the ion
energies is diminished. It is interesting, however, that the plasma DC
potential offset does not appear to change significantly. The data was
reduced to ion energy *vs.* ion mass plots, as given in Figure 9. Obviously,
the reduced mass-dependence of the ion energies at higher pressures is
reflected in a lower calculated neutral plasma temperature. However, as
we begin to sample from a region near the mach disk, the assumption of
a molecular beam flow through the skimmer is unlikely to remain valid. In
fact, with the skimmer at the mach disk we are likely observing a re-
expansion of the plasma from the mach disk. Providing that the density in
the mach disk is sufficiently high that the mean free path is much smaller

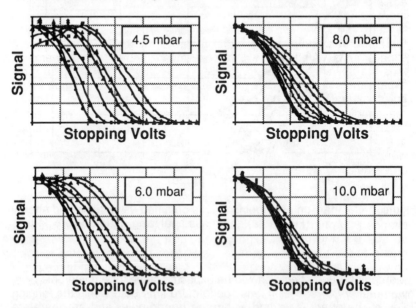

Figure 8 Ion stopping curves as a function of interface pressure.

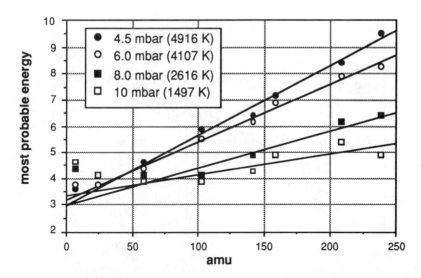

<u>Figure 9</u> Most probable ion kinetic energy *vs.* ion mass as a function of interface pressure.

than the diameter of the skimmer (as it is likely to be), the re-expansion will be of continuum flow. In this case, the plasma expansion will once again be characterized by constant velocity flow, and the interpretation of the data in terms of determining a temperature within the mach disk remains valid. (It might be noted that if the density is sufficiently low that the mean free path is much larger than the skimmer diameter, then the expansion will be effusive, which is characterized by constant energy, and thus the ion energies would collapse to a single value.) Since the re-expansion is likely to be of continuum flow, the results suggest that the bulk temperature near the mach disk is of the order of 1500 K. This is in reasonable agreement with the measurements of Houk[9], who found a mach disk temperature of the order of 2200 K. (In fact, it can safely be assumed that the temperature measurements reported in the present work underestimate the true plasma temperatures by something of the order of 30%; *cf.* the lens bias apparent in Figure 6).

When the nebulizer flow is adjusted, the position of the ionization "bullet" (in the central channel) with respect to the sampler orifice is changed. Higher nebulizer flows cause the channel to move further up, so that the plasma gas extracted through the sampler derives from a region further down towards the load coil, where the plasma is cooler. Lower nebulizer flows cause sampling from a region higher in the plasma where the plasma temperature is higher. Therefore, we would expect that the ion energies will also reflect the change in plasma temperature as the nebulizer flow is varied. Figure 10 provides ion energy *vs.* ion mass plots for nebulizer flows in the range 0.85 to 1.05 l/min. which bracket the top of the nebulizer behaviour plot (the "mountain"). The ion energies do not, in fact, change very much; certainly not enough to require re-optimization of

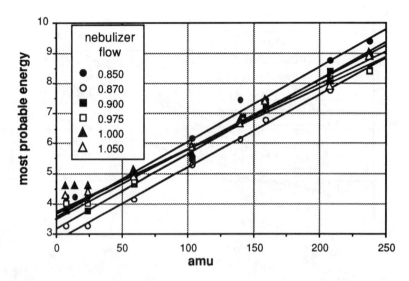

<u>Figure 10</u> Most probable ion kinetic energy vs. ion mass as a function of nebulizer flow.

the lenses (the bandwidth of the lenses is greater than the changes in the ion energies). Over this range of nebulizer flows, the DC plasma potential offset appears to have changed by *ca.* 1 eV. It can be seen, however, that the slope of the plots is a function of the nebulizer flow, resulting in a decreasing calculated plasma temperature as the nebulizer flow is increased. To put these calculated temperatures into perspective, Figure 11 provides the nebulizer parameter plots ("mountain") for Rh^+ (an atomic analyte ion) and for CeO^+ (the oxide ion of one of the most refractory elements). Overlaid upon these curves is the temperature profile determined from ion kinetic energy measurements. It is significant that the analyte ion response maximizes just at the point where the plasma temperature reaches its maximum (or begins to fall, depending on your point of view), and that the oxide ion gains prominence as the temperature decreases (since there is less thermal energy available here to fragment the oxide or its ion).

To this point we have considered primarily atomic ion kinetic energies. Throughout the experiments, the kinetic energy of the CeO^+ ion was also monitored. In every instance, the oxide ion showed a kinetic energy which was lower by 0.5 to 1.5 eV than predicted for an isobaric atomic ion. A typical result, obtained near the "top of the mountain" is given in Figure 12. In this instance, the oxide ion energy is *ca.* 0.64 eV lower than predicted for an isobaric atomic ion of mass 156 amu. Assuming that the DC plasma potential offset was constant (after all, these measurements were obtained simultaneously), use of equation (3) suggests that the CeO^+ ion was, on average, derived from a region in the plasma which was some 761 K cooler than the corresponding atomic analyte ions. The temperature which is calculated here is for the bulk plasma temperature. The implication is that the bulk plasma temperature

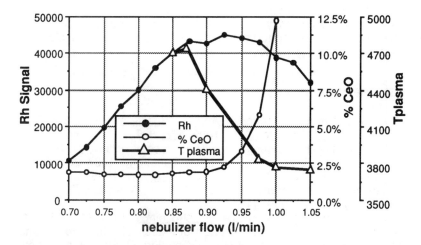

<u>Figure 11</u> Nebulizer parameter plots for Rh⁺ and CeO⁺, together with the temperature profile determined from these ion kinetic energy measurements.

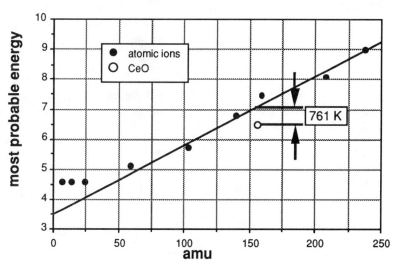

<u>Figure 12</u> Most probable ion kinetic ion energy vs. ion mass plot obtained near the "top of the mountain". The kinetic energy of the oxide ion CeO⁺ is seen to be ca. 0.64 eV lower than that predicted for an isobaric atomic ion.

is, on average, cooler in the region where the oxide ions form. Since the plasma is sampled continuously at a fixed sampling depth, this in turn suggests a temporal variation in the plasma temperature. It is reasonable to assume that the temperature in the micro-region surrounding a large persistent droplet or particle is cooler than in regions which are fully vaporized. In turn, these results suggest that oxide ions are derived from that locally cooler region near a droplet or particle, and that atomic analyte ions are more prominent in regions devoid of these droplets or particles. Work is currently underway in our laboratory to confirm this interpretation using a monodisperse particle generator for which the vaporization point for all particles in the plasma is the same.

REFERENCES

1. D.J. Douglas and J.B. French, J. Anal. At. Spectrom., 1988, 3, 743.
2. S.D. Tanner, Spectrochim. Acta, 1992, 47B, 809.
3. J.A. Olivares and R.S. Houk, Appl. Spectroscopy, 1985, 39, 1070.
4. A.L. Gray and J.G. Williams, J. Anal. At. Spectrom., 1987, 2, 599.
5. J.E. Fulford and D.J. Douglas, Appl. Spectroscopy, 1986, 40, 971.
6. N. Jakubowski, B.J. Raeymaekers, J.A.C. Broekaert and D. Stewer, Spectrochim. Acta, 1989, 44B, 219.
7. D.M. Chambers and G.M. Hieftje, Spectrochim. Acta, 1991, 46B, 761.
8. D.J. Douglas, "Fundamental Aspects of Inductively Coupled Plasma Mass Spectrometry" in "Inductively Coupled Plasma in Analytical Atomic Spectrometry", 2nd edition, eds. A. Montaser and D.W. Golightly, VCH Publishers, New York (in press).
9. H.B. Lim, R.S. Houk, M.C. Edelson and K.P.Carney, J. Anal. At. Spectrom., 1989, 4, 365.

So Where Do We Go From Here?

A. L. Gray
NERC ICP-MS FACILITY, ROYAL HOLLOWAY, UNIVERSITY OF LONDON,
EGHAM, SURREY TW20 0EX, ENGLAND

After a meeting such as this it is useful, before separating to become engulfed in our individual daily problems, to pause for a moment and consider where the accumulated presentations of the week might be pointing for the future. An essential starting point for such speculation must however be an assessment of the current status of the techniques as revealed by the meeting. The majority of the 70 presentations were on ICP-MS but of the 9 on GDMS only 3 were devoted to fundamentals and instrumentation and the remainder to applications. A similar distribution was seen for ICP-MS so that overall 70% of the presentations were devoted to applications of the two techniques and most of these were concerned with routine use in analytical problems rather than exploratory investigation. This shows a very welcome level of assurance and confidence in the methods and suggests that user experience is now sufficiently widespread that problems in their application are regarded as the exception, having been avoided by the adoption, ab initio, of the correct analytical methodology. Such a demonstration of maturity of these powerful but sophisticated techniques is new and this is a most welcome development. In ICP-MS a third of the papers reported work using methods of sample introduction other than conventional solution nebulisation, and 8 of these were on the use of laser ablation as a routine method, some of these reporting highly quantitative analysis. The remaining 7 were divided between flow injection methods and electrothermal volatilisation.

While the applications papers demonstrated new levels of confidence in the use of the techniques a substantial proportion (30%) of the presentations described new instrumental developments which point the way for future evolution. In GDMS a valuable review of the fundamentals of the glow discharge process gave a reference point for new developments in sputtering cells and applications in depth profiling. The use of RF sputtering for non metallic samples suggests a wider range of uses for the technique. Since the bulk of the presentations related to ICP-MS however it was here that new developments offered most of the pointers to future exploitation. Six main areas of development for the immediate or more distant future were described.

At the start of the analytical process sample introduction has great influence on system performance. Since solution samples are the most common much attention is currently paid to alternatives to the conventional, but inefficient, pneumatic nebuliser, with its poor analyte to water vapour ratio in the aerosol. Several different approaches were described which improved efficiency or greatly reduced the amount of water vapour accompanying the analyte or both, including pneumatic or ultrasonic nebulisation with cryogenic desolvation, and the use of high efficiency and direct injection nebulisers. These can reduce polyatomic ion formation which so limits performance below mass 80 and also decrease the significance of refractory oxide peaks. The use of monodisperse droplet generation has also been found to provide 100% sample utilisation and greatly reduced oxide levels even without desolvation. Advances were also reported in laser ablation, particularly in the reduction of ablated spot size to as little as 5 microns. Much of the work reported on sample introduction is capable of immediate application by a wide range of users of existing instruments and so forms the most easily accessible step forward.

Considerable attention is now evidently being paid to the argon ICP used as an ion source. It is perhaps surprising that so little change has been necessary in the conventional form used as an emission source but this is probably due to more significant shortcomings in other areas. Now however more attention is being paid to the causes of instability and noise in the plasma as they contribute to lack of precision in elemental quantitation and isotope ratio determination. Studies of solvent and solution droplet size distribution in the plasma central channel in the processes of atom and ion formation from the analyte and in the effects of additions of molecular gases to the plasma or carrier gas all hold potential for improved performance.

The interface between plasma and mass spectrometer has always been a crucially important stage of the system and studies of the whole ion extraction process from the plasma up to the formation of the ion beam directed into the mass analyser are fundamental to improvement in this area. A number of very interesting investigations were reported which in due time should lead to reduced losses of the extracted ions in transit to the analyser and to better understanding of, and hopefully improvements in mass discrimination and ion suppression.

Until recently most instruments have been equipped with channel electron multiplier ion detectors used in the pulse counting mode and have relied on these to provide the low detection limits for which the technique is noted. Photon sensitivity is a problem with these detectors however and this has usually been dealt with by an axial photon stop in the ion path. This also intercepts a substantial number of ions and thus reduces overall sensitivity and may contribute to mass discrimination. Alternatively the ion path may be directed away from the optic axis and the ions diverted into a detector which is completely screened from photons. This reduces the background count rate to the level of the detector dark rate and thus improves detection limits. Improvement of ion transmission by simplifying the lens system and omitting the photon stop has also enabled Faraday cup detectors to be used which are photon blind while retaining adequate sensitivity for many purposes. The greater simplicity and reliability of such detectors is likely to lead to increased application in the future.

The quadrupole mass analyser has been the almost universal choice for the spectrometer stage, largely because of its natural suitability for the low energy ions produced by the ICP and its relative cheapness. The problems of interferences at the lower end of the mass range can be severe for some analytes however and the greatly improved resolution obtainable from a magnetic sector analyser is often worth the considerable extra cost. Such instruments, which also show very low background levels, give extremely low detection limits even for normal pneumatically nebulised solutions. The determination of Fe, for example, is always a problem on a quadrupole instrument, unless the solvent water is removed from the analyte by cryogenic desolvation or the use of ETV, but on a high resolution magnetic instrument detection limits of less than 10 ppt are obtained, reducing to 10 ppq if cryogenic desolvation is also used. These limits are only achievable in samples in very specialised circumstances and are probably of less general importance than the use of a multi-collector magnetic instrument for isotope ratio determinations reported at the meeting. Precisions equivalent to or better than those obtained from thermal sources are reported and more elements are accessible because of the high source temperature. The greatly simplified sample introduction to the ICP permits a much faster sample throughput and the higher cost compared to thermal, or the less precise quadrupole instruments, is acceptable in some cases. When coupled to a laser ablation stage the capability of high precision isotope ratio determination with microprobe spatial resolution yields a very powerful instrument. Although the argon ICP has been the mainstay of ICP-MS since its inception and only minor additions of other gases have been investigated to improve its performance, it suffers from two unavoidable fundamental limitations. These are that its isotopes and the polyatomic species associated with them, cause severe isobaric interference with some important analyte elements, which prevents or severely restricts their determination, and that its ionisation energy of 15.76 eV is too low for efficient ionisation of elements such as the metalloids and halogens. Of the other rare gases only helium, ionisation energy 24.59 eV, is plentiful enough to form an acceptable alternative but its physical characteristics have so far prevented the development of a satisfactory ICP. A number of low power microwave systems have been built using helium and some coupled to MS systems but it has proved difficult to introduce solution samples to them. The successful development of a medium power annular helium plasma which will accept nebulised solution samples represents a major step forward. Detection limits in the region of 10 ppt for As, Se and Br are reported and coupling to a conventional MS system appears straightforward. Commercial availability of this system will solve a number of difficult analytical problems and it is hoped that this will not be far away.

These six areas of forward development are at varied stages, some, such as high resolution and multicollector instruments, and improved nebuliser systems are commercially available. Alternative ion detectors are already used in some instruments and a helium MIP is close to the market. The other areas are at present important research fields from which may come systems of considerably improved performance in the future. It is just 10 years since a commercial ICP-MS system was first announced by VG Isotopes at the 1982 Vienna IMSC and the degree of maturity shown in the applications reported by users at this meeting is very encouraging. The technique is now fully accepted and in its present form is capable of impressive work. Any enquirer need have no hesitation in employing it at its present status. It is evident however that this is only a step on the way to the fuller realisation of its potential and that these six areas are only the most obvious pointers to the future. The early development of ICP-MS was largely empirical and the greatly increased fundamental effort now being devoted to the more complex aspects such as the plasma itself, the interface and the ion beam forming stages may be expected to yield great rewards in the longer term.

Subject Index